MORE

极致大宅

台湾麦浩斯公司 编

海峡出版发行集团 | 福建科学技术出版社
THE STRAITS PUBLISHING & DISTRIBUTING GROUP | FUJIAN SCIENCE & TECHNOLOGY PUBLISHING HOUSE

目录

极致大宅

坐落于精华地段的都会城堡

A urban castle located in the prime area

garden 庭院

garden 庭院

Picture_TAD ASSOCIATES INC
图片提供_大隐设计

也许可以这么去诠释：家就是财富的一个展示场所。本案为位于台北精华地段的豪宅，面积之大即便是郊区大宅也只能望其项背。屋主期望设计师为他们量身设计上下两层的居住空间，每个区块都有其各自的功能，如城堡般完美无缺。

因应屋主好客的个性，一楼规划为主要的公共空间，玄关入口的长廊可依时序摆放不同的艺术作品，以小型画廊的概念，将美术馆功能置入于居家空间。客厅分为两个区块，可容纳20人的主客厅让访客可各自聚会，而副客厅让屋主在举办小型音乐会时无需担心场地的布置问题，宽敞与舒适的定义在此处全然由于空间的尺度而又有不同的意义。在户外，规划了私人庭园与游泳池，朱铭与其他艺术家的雕塑，则提升了庭院的艺术性。

圆形语汇为主的餐厅设计，可看出屋主其重视东方文化的内敛个性，讲求和谐与低调的意涵延伸至空间设计中，空间让位给材质、家具与艺术品，摒除多余装饰的设计思想更能衬托屋主的地位与背景。

We may interpret a home as a place for displaying power. The project is a mansion in Taipei City' s prime area. The total area of the mansion is even bigger than a villa in the suburban area. The owner authorizes the designer to design the living areas of the two floors. And all of the spaces have their own functions and therefore work together like a perfect castle.

The first floor is the public area because owner likes guests. The hallway of the entrance can displays different artworks with different seasons like a little gallery at home. The living room is divided into two districts and redefines the idea of spaciousness and coziness. The first living room can accommodate a gathering of twenty guests. The owner can hold a concert in the second living room. In the outdoor area, the owner has a private garden and swimming pool. Ming Ju and other designers' sculptures create the artistic atmosphere for the yard.

Circle is the theme of the kitchen; it shows the owner' s reserved personality and his love for oriental culture. The ideas of harmony and low profile have extended to the space design. Furniture and artworks stand out in the space. Simple style actually sets off the owner' s rank and background.

空间案名　帝宝私宅
设计者　王镇华
设计公司　大隐设计集团
空间地点　台北
空间性质　电梯大楼
空间面积　1500 ㎡
空间格局　玄关、主客厅、次客厅、厨房、餐厅、水疗室、主卧室、主卧更衣室、主卧浴室、小孩房、客房、室内泳池、户外泳池
主要建材　进口石材、镀钛金属件、染色橡木板、柚木、紫檀木、白金箔、羊毛地毯、皮革、壁布

Title / Private residence in Depo
Designer / Arthur Wang
Design company / TAD ASSOCIATES INC.
Location / Taipei City
Category / Elevator building
Area / 1500 m^2
Layout / Entrance, main living room, additional living room, kitchen, dining room, SPA room, master bedroom, master dressing room, master,bathroom child's rooms, guest、kid's room, guest room, indoor swimming pool, outdoor swimming pool
Building materials / imported stone, titanium-plated metallic components, dyed oak wood boards, teak wood, red sandalwood, platinum foil, wool carpet, leather, wallpa

01_玄关　02_客厅　03_客厅　04_客厅　05_客厅
01_hallway　02_living room　03_living room　04_living room
05_living room

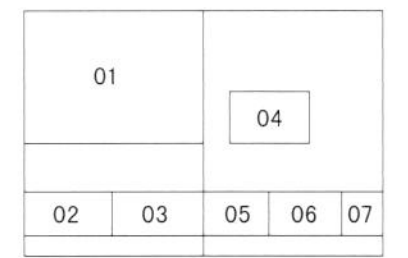

01 _ 餐厅　02 _ 书房　03 _ 主卧室　04 _ 厨房　05 _ 室内温水泳池　06 _ 主卧浴室　07 _ 梯厅
01 _ dining room　02 _ study room　03 _ master bedroom
04 _ kitchen　05 _ indoor swimming pool　06 _ master bathroom　07 _ ladder hall

Designer Data

王镇华
现　任/
大隐设计集团台北、上海主持人
学　历/
辅仁大学毕业
经　历/
1985~1989 珠宝设计师
1999 罗芙奥拍卖公司、睿芙奥艺术顾问公司
简　介/
王镇华从事室内设计领域超过20年，在这期间陆续成立了A&F设计事务所、大隐室内设计公司(TAD ASSOCIATES INC.)，并将事业领域扩展至中国内地。王镇华以其对艺术品长期涉猎的经验，为委托客户提供艺术品陈设及收藏的顾问。目前其所主持的设计公司主要以承接私人豪宅以及企业招待所、办公空间为主。

Arthur Wang
Principal of TAD ASSOCIATES INC., Taipei and TAD ASSOCIATES INC., Shanghai
Education
Fu Jen Catholic University
Professional
1985–1989 Jewelry designer of ALL GEMS JEWELRY
1999 Ravenel Art Group
Brief bio
Arthur Wang has been in the interior design for more than twenty years. He founded A&F INC and TAD ASSOCIATES INC and has expanded his business to mainland China. Wang is an artworks counselor for his clients. His companies focus on the interior design service for private mansions, enterprise guest houses and offices.

拥抱山林与湖泊的居所

Place that embraces mountain and lake

two building balconies 主栋二楼阳台

Construction outward appearance 建筑物外观

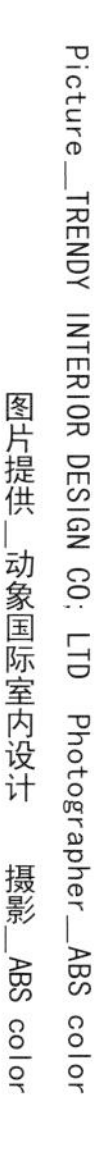

退休后的生活，不再在都市的尘嚣中终日忙忙碌碌，而是返回大自然的怀抱，享受风和日丽下的恬静。屋主原本仅是征询设计师的意见，期望未来的居住场所能充分地满足生活所需，后来经过设计师的细腻规划，原已开始动工的房子有了不一样的样貌。

本案分为主客两栋，整个基地由一个人工湖所环绕。设计师刻意将建筑立面的开窗加大，强调了室内与户外的连接关系，将绿意与阳光带入室内，营造出宽敞辽阔的视野。而精简的设计，赋予空间宽阔感与自由不同的意义，也强调了居住空间与自然环境的关系，进入房子就能充分体会到什么才叫"过生活"。

考虑到屋主好客，客栋主要作为招待访客之用，除了设计套房式的空间，三面皆落地窗的餐厅，成了访客最喜爱的场所。

说奢华，再大的房子也比不过拥有一座山与一面湖的景致，设计的精简，在此处让大与奢华有了完全不同的诠释。

The owner decides to shun away urban life and goes back to nature after retirement. Originally the owner just counsels the designer and only expects his house to fulfill its basic needs. However, the designer reshapes the house through delicate planning.

The villa has main building and addition. The foundation is surrounded by an artificial area. The designer deliberately enlarges the area of the window to enhance the connection between indoor and outdoor areas, bring the greenness and sunlight to the room and create a broad view. And the concession here redefines the spaciousness and freedom of a pace. The simple design language emphasizes the relationship between living space and the natural environment. One will never know what "living a life" means until he/she enters the house.

The main building is mainly for hosting guests because the owner is hospitable. All of the rooms are suites. The most frequented spot is the dining room with three French windows.

The most luxury of the house will never surpass beauty of the scenery of a mountain and a lake. The simple design reinterprets the bigness and luxury here.

House Data

空间案名　屏东大原邸
设 计 者　谭精忠
参与设计　杨舒如、许玉臻、林青蓉 、黄利画
设计公司　谭精忠室内设计工作室
空间地点　屏东县
空间性质　庄园别墅
空间面积　主栋671 m² 、客栋307 m²
空间格局　玄关、客厅、厨房、餐厅、客浴、主卧室、主卧浴室、书房
主要建材　木皮、石材、镀钛金属件、烟熏木地板、喷漆

Title / The house in Da-yuan, Ping-dong
Designer / David Tan
Assistant designer / Shu-ru Yang, Jane Hsu, Josephie Lin, Peter Huang
Design company / TAN CHING-CHUNG STUDIO
Location / Ping-dong County
Category / Villa
Area / main building 671m², addition 307m²
Layout / entrance, living room, kitchen, dining room, guest bathroom, master bedroom, master bathroom, and study
Building materials / Grain Brushing, stone, titanium, smoked wood flooring, and spray painting

01 02 03 04 06 05 07

01 _ 客栋餐厅与厨房　02 _ 主栋露台　03 _ 主栋二楼过道　04 _ 主栋餐厅
05 _ 主栋次起居室　06 _ 主栋起居室　07 _ 主栋餐厅
01 _ dining room and living room　02 _ main balcony　03 _ main two building corridor
04 _ dining room　05 _ reception room　06 _ reception room　07 _ dining room

01 _ 主栋书房　02 _ 主栋主卧室　03 _ 客栋餐厅　04 _ 客栋一楼客房　05 _ 主栋主卧浴室　06 _ 客栋二楼露台
01 _ study room　02 _ master bedroom　03 _ dining room
04 _ guest room　05 _ master bathroom　06 _ two building flat roofs

Designer Data

谭精忠

现　任/

动象国际室内设计、大隐设计集团主持人

学　历/

复兴美术工艺科毕业

经　历/

1984　任职于谭国良设计事务所

1989　成立谭精忠室内设计工作室

1999　成立动象国际室内设计有限公司

2002~2003 成立大隐室内设计有限公司上海、台北分公司

简　介/

1984年毕业之后即跨入室内设计界，1990年获意大利TECNHOTEL PACE室内设计奖优胜奖。2006、2007年连续获得亚太区室内设计大奖、日本JCD以及TID台湾室内设计大奖等多个奖项。2007年评为十大样板房设计师之一。

David Tan

Principal of TRENDY INTERIOR DESIGN CO, LTD and TAD ASSOCIATES INC

Professional

1984　Worked at Tan Guo-liang Studio

1989　Established TAN CHING-CHUNG STUDIO

1999　Established TRENDY INTERIOR DESIGN CO, LTD

2002-2003 Establish TAD ASSOCIATES INC., Taipei and TAD ASSOCIATES INC., Shanghai

Education

Fu-Hsin Arts School

Brief bio

David Tan has entered interior design industry since his graduation in 1984. He gained TECNHOTEL PACE INTERIOR DESIGN 1990 AWARDS. He also gained Asia Pacific Interior Design Awards, JCD Design Awards (Japan) and TID Award (Taiwan) in 2006-2007. Tan received the China Top 10 Sample House/Room Designer Awards in 2007.

让空间成为光线的画布

Turning space into the light's canvas

living room 客厅

courtyard 庭院

重新赋予旧屋以新的面貌是本案设计的最大挑战。在注重经济、环保的前提下，设计师保留了大部分既有的建筑本体。外墙刻意使用的黑色外挂砖，撷取艺术家罗斯柯（Rothko）的巨幅直式画布概念。而散落的建筑立面开窗，通过不同大小的长方形色块平行排列整合，拉近了人与建筑的距离。

光线在本案中成为极重要的角色。客厅的挑高设计，不但展现出大宅的宽敞气势，留白的主墙更是光线的画布。随着时间的流动，墙面的风景也不断地改变，让人深刻感受到时光的变换。而在户外，车库上木制、铝制的格栅线条与风铃木相互竞影，阳光也在铝格栅露台上游移，增添了多样表情。

尺度在本案中已不是问题，借助空间的释放，空间拥有了多样面貌，也让人有了感性的体验，因而能为生活创造更多的趣味性。即使像车库车道的长向直线设计，也仿佛营造出飞机跑道的印象，尤其当走道尽头天井上方的阳光洒下，不但一改车库晦暗的印象，还赋予一种特有的宁静感。

The project's main challenge is to give an old house a new refreshing appearance. We were given the task of preserving the original architecture under the limits of economic constraints and the demands for an environmentally friendly result. The outer walls have a deliberate design of secured black brick facing, combined with the concepts of large vertical canvas inspired by designer Rothko. The scattered vertical sliding sash windows are connected visually by rectangular color blocks arrayed in a parallel formation to close the distance between the human user and the architecture.

Light plays an extremely important role in this project. The increased ceiling space of the living room gives an open and generous atmosphere appropriate for such a large villa. The white walls act as a canvas for light, with the wall landscape changing throughout the day, giving the residents a feel of the ebbs and flow of time. Externally, the wooden and aluminum construction on the garage competes with the golden trumpet trees for light and shadow, giving more character to the canvas on the wall as the sunrays flow across the newly erected aluminum fences.

Scale is not an issue in this project. The freedom to operate spatially allows the designers to give the compartments a diverse array of features, giving a sensational experience and a livelier variety in residence. The driveway to the garage is a long straight path, giving a sense of velocity not unlike a landing aircraft. As the car reaches the end of its journey into the garage, the skylight window bathes it in shimmering rays, overturning the shadowy darkness of other traditional garages whilst imparting a soothing calm to speed.

Pictures_Waterfrom Design Photographs_Sam+Yvonne
图片提供_水相设计 摄影_Sam+Yvonne

House Data

空间案名　向罗斯柯致敬
设 计 者　李智翔
参与设计　陈凯伦、黄培伦
设计公司　水相设计
空间地点　台北市
空间性质　别墅
空间面积　932 ㎡
空间格局　四房三厅、两厨房、车库、六卫
主要建材　木纹水泥砖、钻石檀木、柚木、集成木地板、染色栓木木皮、风化木板、印度黑大理石、雪白碧玉大理石、铁件、抛光铝材

Project name / An Homage to Rothko
Designer / Li Zhi-xiang
Participants / Kai-Lun Chen, Pei-Lun Huang
Company / Waterfrom Design
Waterfrom DesignLocation / Taipei
Type of space / Villa
Area / 932 m^2
Layout / 4 bedrooms and 3 living rooms; 2 kitchens, garage, 6 auxiliary rooms
Main materials / Wood-grained concrete tiles, betis wood, teak, plywood floor, ash wood dyeing, weathered board, Indian black marble, white marble, steel components, anodized aluminum

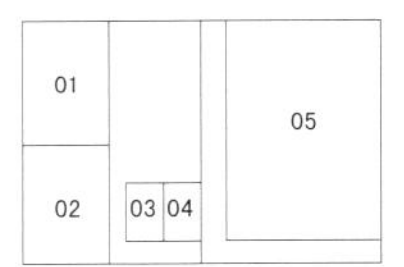

01 _ 玄关　02 _ 客厅　03 _ 庭院　04 _ 建筑外观　05 _ 客厅
01 _ font porch　02 _ living room
03 _ courtyard　04 _ external view　05 _ living room

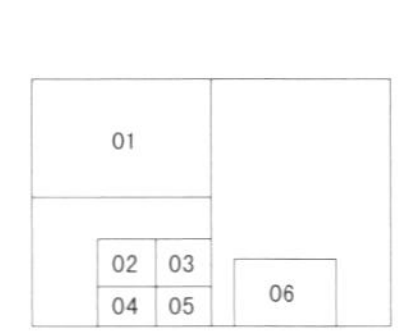

01 _ 客厅与餐厅　02 _ 客厅　03 _ 卧室
04 _ 餐厅与厨房　05 _ 车库入口　06 _ 车库
01 _ living room and dining room
02 _ living room 03 _ bedroom
04 _ dining room and kitchen 05 _ garage entrance
06 _ garage

1F

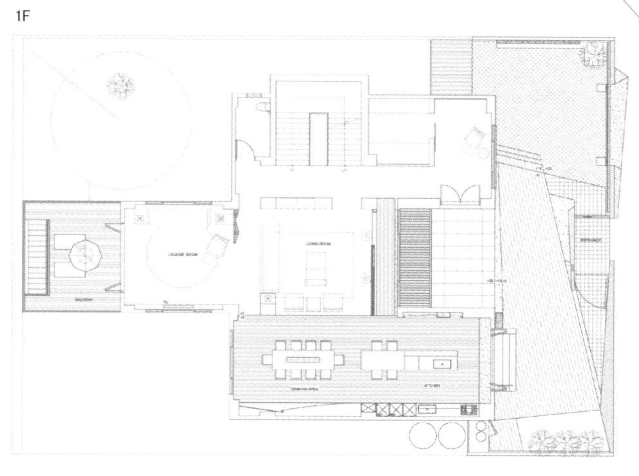

2F

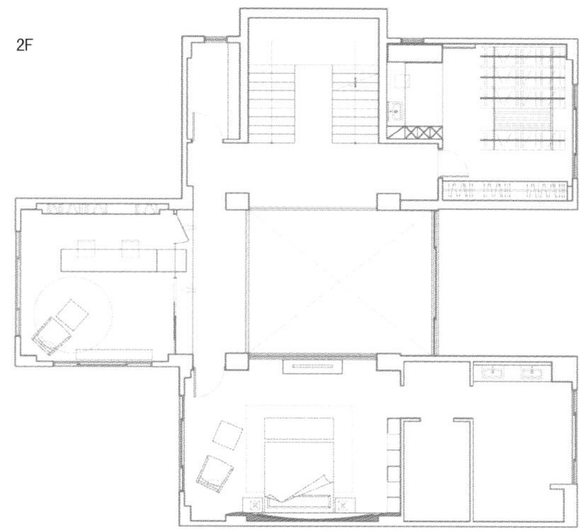

B1

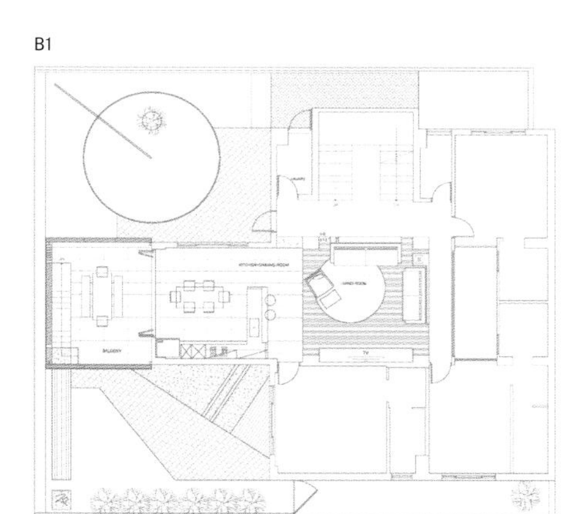

B2F

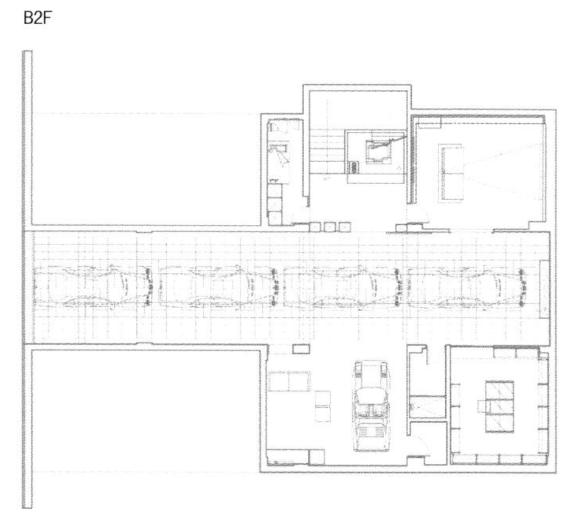

Designer Data

李智翔
现　任/
水相设计　设计总监
学　历/
1998~1999 丹麦哥本哈根大学研修毕业
1997~2000 美国纽约布拉特学院室内设计硕士
经　历/
2008　　　水相设计 设计总监
2004~2008 荷果设计 设计总监
2001~2003 MMoser 香港商穆氏设计集团
2000~2001 李玮珉建筑师事务所
简　介/
重新再定义空间，创造真正量身订制的个性住宅，富含幽默感的设计语汇，是李智翔设计师不断尝试在空间注入的特质。近期作品更追求“自然光与线条比例”的相对关系，以更内敛但不含蓄的手法处理每一项细节，创造出独特的设计风格。

Li Zhi-xiang
Waterfrom Design, Design director
Education
1998–1999　The University of Copenhagen, Research programmes
1997–2000　Pratt Institute, New York, USA. Mater of Interior Design
Professional
2008–present　Waterfrom Design, Design director,
2004–2008　Hercore Design, Design director,
2001–2003　MMoser
2000–2001　LWMA
Brief bio
Li Zhi-xiang attempts to instill the space with the refreshing quality by redefining the space, creating the tailor-made personal houses, and employing humorous vocabulary. His recent works pursue the relative relationship between "natural light and linear proportion." And Li handles every detail in a restrained but not reserved way, in order to create absolutely unique designs.

浪漫古典大宅绽放唯美情怀

Aesthetics in a romantic and classical mansion

entrance corridor 玄关廊道

stairway entrance 楼梯入口

当古典遇上了美式，经典中更能透出精彩表现！本案空间整体以欧式古典浪漫风格为主，搭配时尚现代的美式元素，为喜爱唯美风格的屋主夫妻量身打造精品级豪宅。

由于女主人喜爱芬芳优雅的花朵，因此本案设计的重要语言便是花朵图案。设计师运用新古典元素，如花朵图案刺绣、壁纸，以及楼梯的镂空藤蔓铁艺扶手等等，搭配色调淡雅的地板与具有金属质感的装饰壁纸，充分展现出大宅耀眼且华丽的气势，同时也让空间呈现出多彩多姿的典雅氛围。

另外，水晶元素的运用亦为本案重点之一。由于建筑物本身具有相当壮观的挑高优势，搭配水晶灯更能展现豪宅的宽敞，因此屋主特别指定选用施华洛世奇品牌的水晶灯饰，营造出大宅的风范。楼梯处的巨型垂挂式水晶灯高约7m，由三楼天井垂降至一楼，在古典风格的居家中注入了奢华的气度。而水晶灯饰也出现在各处端景中，如主卧室与视听室中间的廊道，即运用水晶吊灯与金色花朵图案壁纸打造出空间端景，带来细腻而华美的视觉感受。

When classical meets American style, vintage and modernity are both at hand! The property itself is decorated with European classical and romantic style as the base, in addition to a hint of contemporary American flair. Through this unique combination, this house has been tailored made for the house owners to suit their preferred aesthetic sense.

Due to the fact that the hostess has a liking for flowers, the main concept of this design is flowering pattern. The design has applied many neo-classical design elements, such as embroidery, the use of flower-patterned wallpapers as well as using staircases with iron handrails in which flower and vine patterns are carved on. This sort of design complements well with the light-colored flooring as well as wall decorations with a metallic feel. Through these arrangements, the dazzling and elegant nature of the apartment is fully revealed.

Furthermore, the use of crystals is another important aspect of this design. Since the building itself has a considerable advantage of being endowed with a particularly high ceiling, the use of chandelier can further emphasize the expansiveness of the layout. Furthermore, the house owners specially choose the Swarovski-made chandelier, to bring a sense of sophistication to the house design. This gigantic chandelier is hanged from the ceiling next to the stairs, and has a length of about 7 meters. The ceiling is of three stories high, and this chandelier hangs down all the way from the 3rd floor to the 1st floor. A sense of luxury has been created through the use of classical styles. Furthermore, chandeliers are used across different locations within the villa, for example, on top of the corridor in between the master bedroom and the audio-visual room, the use of chandelier and the golden-flower patterned wallpaper both help to create a visual experience that is both delicate and luxurious.

Pictures_D-A design Ltd.
图片提供_艺念集私

空间案名 桃园 美式别墅
设 计 者 黄千祝、张绍华
设计公司 艺念集私空间设计有限公司
空间地点 桃园县
空间性质 别墅
空间面积 891 m²
空间格局 1楼：玄关、客厅、餐厅、开放式厨房、透明红酒间、客房、佣人房、工作间、储藏间、车库。2楼：视听室与书房、主卧房、更衣间、主卧浴室。3楼：起居厅、二间小孩房、各有更衣室与浴室、书房
主要建材 玄武岩＋喷漆、进口壁纸、大理石、复古地砖、立体花砖、实木地板、锻铁扶手、水晶灯、纱帘、玻璃、不锈钢材、仿鳄鱼皮绷皮、发光二极管灯、录音室专用吸音建材

Case Name / Taoyuan American-styled villa
Designers / Anita Hwang & David Chung
Company / D-A design Ltd.
Location / Taoyuan County
Nature of the apartment / villa
Area / 891 m²
Outline / 1F entrance,lounge,dining room, open-designed kitchen, transparent red wine cabinet, guest's room, servant's room, work space, storage, garage;2F audio-visual room &study room, master bedroom, dressing room, master bathroom;3F living room, two kid's rooms with individually built dressing room and bathroom, study room
Main building material / Basalt and spray-paint, imported wallpaper, marble, retro-styled tile, stereoscopic patterned glaze tiles, wood floors, iron-railed handle, crystal lamp, sheer curtain, glass, stainless steel, tightened crocodile skin,LED light, sound absorbing materials used at audio-visual room.

01 _ 餐厅 02 _ 餐厅红酒室 03 _ 视听室
04 _ 餐厅 05 _ 客厅
01 _ dining room 02 _ dining room, wine room
03 _ auditorium 04 _ dining room 05 _ living room

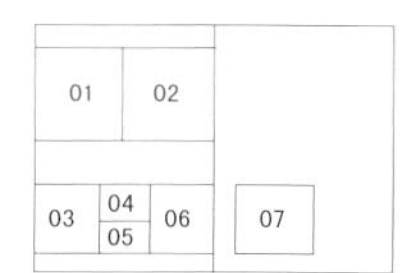

01 _ 客厅　02 _ 主卧　03 _ 廊道端景　04 _ 卫浴　05 _ 更衣室　06 _ 卫浴　07 _ 卫浴
01 _ living room　02 _ master bedroom　03 _ corridor side view
04 _ bathroom　05 _ dressing room　06 _ bathroom　07 _ bathroom

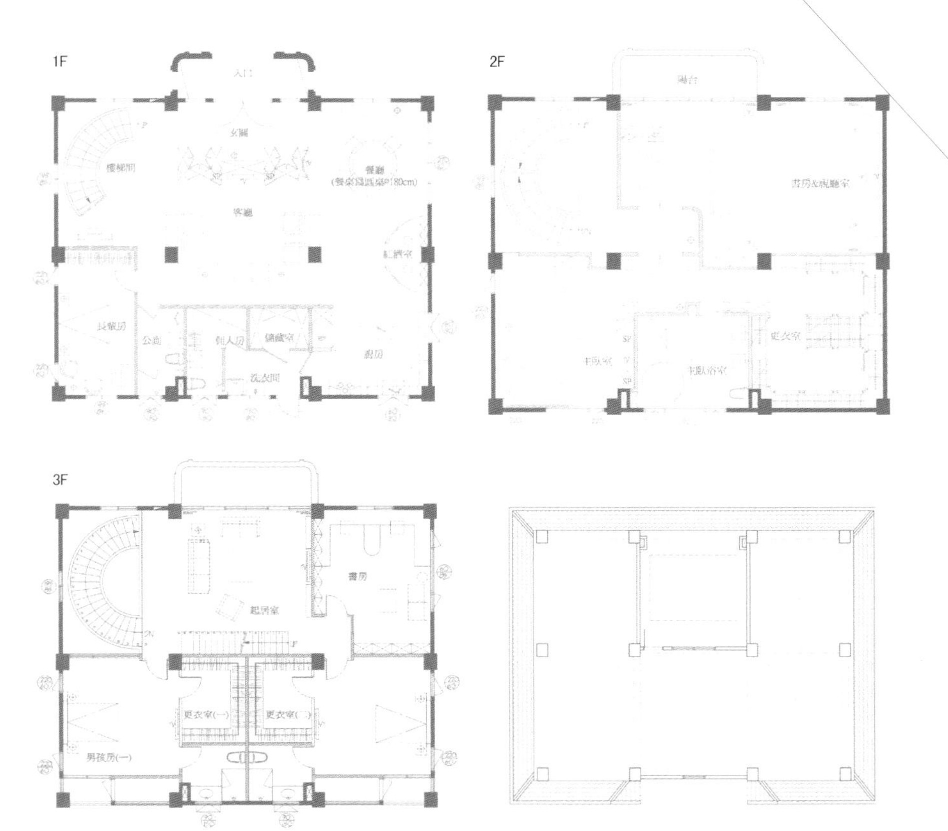

黄千祝

现　任／

艺念集私空间设计有限公司 创意总监

学　历／

协和工商职业学校室内设计科毕业，美国普莱斯顿大学高级工商管理硕士

张绍华

现　任／

艺念集私空间设计有限公司　执行总监

学　历／

复兴美工空间设计专业毕业

简　介／

从不设限自我风格，善用素材本身的美感，利用不同材质及配色展现空间变化。擅长使用建材的特性，特别专注于铸铁材质的运用，将其自然呈现于空间设计中。创意的混搭装置美学，营造出具有个人特殊风格及时尚的空间，让居住者得以释放压力，又呈现出空间的无限可能性。

Anita Hwang

Current Position

D–A design Ltd. Chief Creative Officer

Education

Unit of Interior Design, Xie–He High School of Industry and Commerce, Preston University / EMBA

David Chung

Current Position

D–A design Ltd. Chief Executive Officer

Education

Graduated in Division of Space Design, Unit of Art and Craft, Fu–Hsin Trade and Art School

Brief bio

Never limit oneself to a certain style, we are able to show off the natural beauty of the materials themselves. By using different materials as well as color complementation, the variation in the spatial design is fully exhibited. We are also good at taking advantage of the special properties of the materials themselves. A special emphasis is placed on the use of casted irons, and naturally presents them in our designs. Finally, a fashionable space is created where unique and personalized styles are valued by mixing creativity and aesthetics. This allows the inhabitants to release the tension in their lives as well as creating designs with infinite possibilities.

现代大宅尽显东方人文情调

Modern mansion appears oriental humane mood

living room 客厅

greenhouse 花房

结合西方现代精神以及东方人文底蕴，设计师将不同的元素融合，让居家展现出时髦且温柔敦厚的气质。

如何将大宅的气度显现，是设计师的设计重点。公共空间仅通过玄关的转折，以小见大的设计手法，带出空间的宽广。艺术品的摆设让玄关透出不凡品位。客餐厅至书房则采用长方形规划，不完全隔断的设计，让视线可沿着大面积的玻璃直至书房，而充足的采光更加强空间的宽阔感。主客厅设计讲求宽敞与奢华质感，主墙采用柚木片营造出自然气息，放松而舒适的设计体贴访客，也低调地诉说屋主的个性。而客餐厅间以隔屏区隔，利用架高的木地板串联，动线与视野的开放再度强调出空间的宽敞。

在看似西化的设计中，实则蕴含东方内敛的情调。靠窗处的造景以及利用老木头作为植物容器的手法，都将国画中造境的艺术语汇转化为空间设计。人文气质的强调，也将屋主打坐的习惯加以考虑，满足动中求静的心情。主卧空间的设计，满足女主人对于奢华的期待，对开式的浴室门片，打造饭店般的情境，让沐浴成为生活中的顶级享受。

The designer combines different elements of western modern spirit and eastern humane mood and makes the home displays both modern and gentle manner.

The designer's main task is to display a mansion's loftiness. The public area is seen through the entrance; the spiral design introduces the spaciousness of the place. The array of the artworks reveals the owner's extraordinary class. The dining room, living room and the study are shaped like rectangular. The design of incomplete partition allows the vision to reach to the study along with the large glass. Moreover, sufficient light enhances the spaciousness of the area. The living room focuses on sense of airiness and luxury. The main wall is made of teak wood chip in order to create the natural sense. Relaxing and cozy design makes the visitors feel at home and silently expresses the owner's taste. The dining room and the living room are divided by a partition and connect with the elevated wooden floor. The openness of moving line and vision once again emphasize the airiness of the place.

The seemingly westernized design actually contains eastern reserved mood. The landscaping by the windows and flower pots made of old wood actually turn the elements of Chinese painting into the space design. The designer also takes the owner's meditation habit into account to meet the owner's desire for quietness in the fast-paced world. As for the design of the master bedroom, the designer tries to satisfy the mistress's expectations for luxury. Additionally, the bathroom with off-style sliding doors makes the room like a hotel suite, allowing bathing become the top luxury in life.

Picture_TAD ASSOCIATES INC.
图片提供_大隐设计

House Data

空间案名	慕夏四季
设 计 者	王镇华
设计公司	大隐设计集团
空间地点	台北
空间性质	电梯大楼
空间面积	628 m²
空间格局	玄关、客厅、起居室、轻食餐厅、重食厨房、工作间、储藏室、厨房、书房×2、花房、主卧室、主卧更衣室、主卧浴室、小孩房×3、视听室、客浴、管家房
主要建材	石材、染灰白橡木木皮、进口壁布、白橡木地板、茶色镜

Title / Mucha's Four Seasons
Designer / Arthur Wang
Design company / TAD ASSOCIATES INC.
Location / Taipei City
Category / Elevator building
Area / 628m²
Layout / Entrance, living room, lounge, light food dining room, heavy meal kitchen, studio, kitchen, study x2, greenhouse, master bedroom, master dressing room, kids room x3, audio-visual room, guest bathroom, housekeeper's room
Building materials / tea-colored mirrors, teak flooring, dyed white cherry bark, leather, imported wallpapers

01 03
02 04

01 _ 玄关 02 _ 书房 03 _ 起居室 04 _ 花房
01 _ hallway 02 _ studyroom 03 _ reception room 04 _ greenhouse

01	02	03	
04		05	06 / 07

01 _ 花房与餐厅　02 _ 起居室　03 _ 书房　04 _ 餐厅
05 _ 卧房　06 _ 主卧浴室　07 _ 主卧浴室
01 _ greenhouse & dining room　02 _ reception room　03 _ study room
04 _ dining room　05 _ bedroom　06 _ master bathroom　07 _ master bathroom

Designer Data

王镇华
现　任／
大隐设计集团台北、上海主持人
学　历／
辅仁大学毕业
经　历／
1985~1989 珠宝设计师
1999　罗芙奥拍卖公司、睿芙奥艺术顾问公司
简　介／
王镇华从事室内设计领域超过20年，在这期间陆续成立了A&F设计事务所、大隐室内设计公司(TAD ASSOCIATES INC.)，并将事业领域扩展至中国内地。王镇华以其对艺术品长期涉猎的经验，为委托客户提供艺术品陈设及收藏的顾问。目前其所主持的设计公司主要以承接私人豪宅以及企业招待所、办公空间为主。

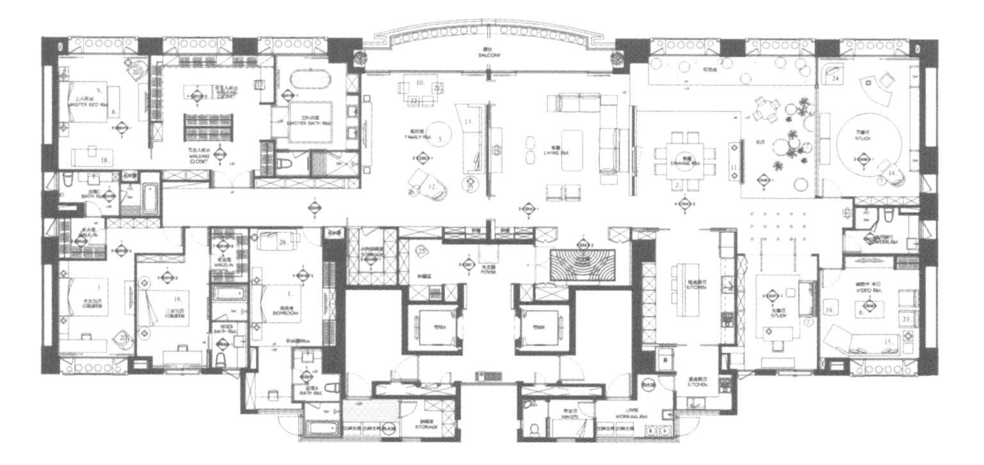

Arthur Wang
Principal of TAD ASSOCIATES INC., Taipei and TAD ASSOCIATES INC., Shanghai
Education
Fu Jen Catholic University
Professional
1985–1989　Jewelry designer of ALL GEMS JEWELRY
1999　Ravenel Art Group
Brief bio
Arthur Wang has been in the interior design for more than twenty years. He founded A&F INC and TAD ASSOCIATES INC and has expanded his business to mainland China. Wang is an artworks counselor for his clients. His companies focus on the interior design service for private mansions, enterprise guest houses and offices.

hallway 玄关

给自己一个宁静的生活空间

Grant yourself a serene living place

living room & dining room 客厅与餐厅

living room 客厅

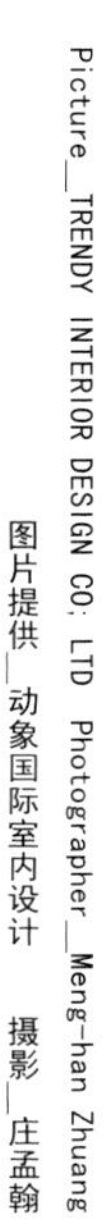

屋主从小在台北市天母地区成长，对此地怀有一份难以割舍的情感，因此在事业有成、人生将迈入下一阶段的分际，买下这间合并屋的大宅，作为给自己与家人的一份大礼。

屋主个性好客，但又需要一个可以尽情放松、独享宁静的空间。考虑到屋主在心理层面与实际功能的需要，故空间规划为公共空间与私人空间两个部分，且两空间有所区隔。公共空间的设计融合招待所的概念，餐厅与客厅自成一个区块，垂直的中轴线将两个空间整合在一起，旁边则邻接女主人的工作间。考虑到屋主的大儿子已成年，因此将其卧室规划在此区域，让小主人也可拥有自己的独立空间。

私人区域以宽敞、宁静作为主要的设计考虑。主卧的更衣室通过梳妆台高墙区隔，当季衣柜与换季衣柜的设置，让男女主人可视需求将衣物分类收纳。主浴室可看到阳明山的景观，成为屋主充分放松的休憩空间。

The owner was growing up in Tian-mu, Taipei, so he has strong attachments to this place. As he enjoys his career achievements and is going to march into the next step of his life, he decided to buy a combined condo as a big present for himself and his family.

The owner is hospitable yet needs a place to relax, restore and release his energy. The designer therefore divides the house into public and private areas to meet the owner's psychological and tangible demand. The public area integrates the idea of hostel, the dining room and restaurant become an independent block; the vertical axis integrates two spaces, next to the mistress's garden. The owner's elder son is a grown-up; therefore his room is located in the public area, so that he can own his personal space.

The themes of the private area are spaciousness and serenity. The dressing room in the master bedroom is segmented by the wall of the dressing table. The designer designs the wardrobes of the current season and the last seasons, so that the owner and his wife can classify their clothes based on their need. The couple can also see the view of Yangming Mountain from the master bathroom. Overall, the private area is indeed a place for relax.

01	03
02	04

01 _ 工作花房　02 _ 客厅　03 _ 餐厅　04 _ 起居室
01 _ greenhouse　02 _ living room　03 _ dining room　04 _ reception room

House Data

空间案名　仰哲韩公馆
设 计 者　谭精忠
参与设计　翟骏、陈敏媛
设计公司　谭精忠室内设计工作室
空间地点　台北市
空间性质　电梯大楼
空间面积　519 m²
空间格局　玄关、客厅、厨房、热炒区、书房、主卧室、主卧浴室、主卧更衣室、客浴、小孩房×2
主要建材　染色与喷漆橡木皮、镀钛金属件、皮革

Chateau Mansfield, Han Residence
Designer / David Tan
Assistant designer / Jun Zhai, Min-yuan Chen
Design company / TAN CHING-CHUNG STUDIO
Location / Taipei City
Category / Elevator building
Area / 519m²
Layout / entrance, living room, kitchen, fried area, study, master bedroom, master bathroom, master dressing room, guest bathroom, children's room x2
Building material / stained oak, titanium alloy, leather wall

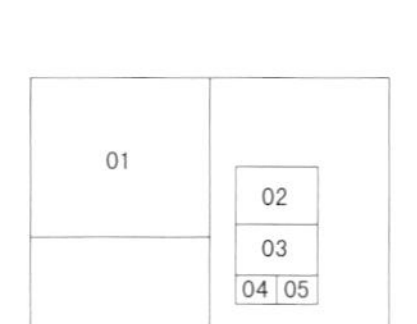

01 _ 主卧室　02 _ 主卧更衣室　03 _ 书房
04 _ 浴室　05 _ 主卧浴室
01 _ master bedroom　02 _ dressing room
03 _ study room　04 _ bathroom
05 _ master bathroom

Designer Data

谭精忠
现　任／
动象国际室内设计、大隐设计集团主持人
学　历／
复兴美术工艺科毕业
经　历／
1984　　任职于谭国良设计事务所
1989　　成立谭精忠室内设计工作室
1999　　成立动象国际室内设计有限公司
2002~2003 成立大隐室内设计有限公司上海、台北分公司
简　介／
1984年毕业之后即跨入室内设计界，1990年获意大利TECNHOTEL PACE室内设计奖优胜奖。2006、2007年连续获得亚太区室内设计大奖、日本JCD以及TID台湾室内设计大奖等多个奖项。2007年评为十大样板房设计师之一。

David Tan
Principal of TRENDY INTERIOR DESIGN CO, LTD and TAD ASSOCIATES INC
Professional
1984　Worked at Tan Guo-liang Studio
1989　Established TAN CHING-CHUNG STUDIO
1999　Established TRENDY INTERIOR DESIGN CO, LTD
2002-2003　Establish TAD ASSOCIATES INC., Taipei and TAD ASSO-CIATES INC., Shanghai
Education
Fu-Hsin Arts School
Brief bio
David Tan has entered interior design industry since his graduation in 1984. He gained TECNHOTEL PACE INTERIOR DESIGN AWARDS in 1990 . He also gained Asia Pacific Interior Design Awards, JCD Design Awards (Japan) and TID Award (Taiwan) in 2006-2007. Tan received the China Top 10 Sample House/Room Designer Awards in 2007.

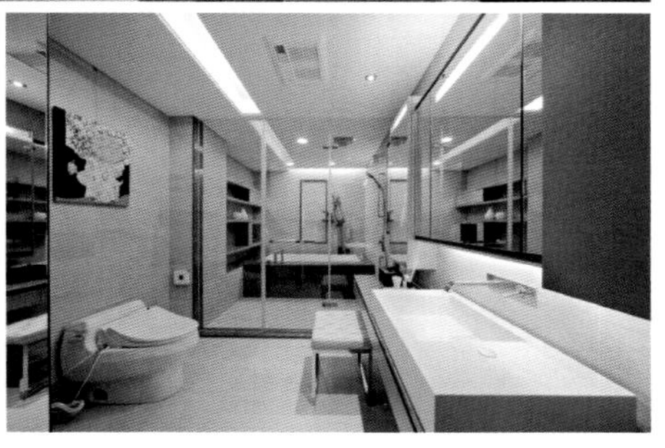

留白，千金难买的奢华

Leave it blank, a luxury even gold can't buy

dining room 餐厅

1F corridor, dining room 1楼廊道、餐厅

干净的背景，没有任何多余的装饰，是这个大宅给人的第一印象。设计师摒除过多设计语汇所带来的干扰，将空间的特色还原至最纯净的状态。也因为隐藏在背后的细腻与内敛，看不到杂乱的立面与复杂的切割线，因而让空间能以最自由的状态呈现。

以楼梯作为空间的中介，一楼玄关的左方为客厅，右方则是餐厅。无任何视觉阻碍的设计，在自然光线的装点下，更能展现出空间的洁净与宽敞。线条是空间设计中的主要元素，但设计师更重视墙面的整体表现，空间合宜的比例才是设计的重点。经过计算的长宽高比例，展现出设计的精准，而表面材质如石材、铁件与进口石英砖，又能以其自身的优异触感，带来感官上的享受，让空间自然呈现出不凡的品位。

在空间的设计上，设计师利用拉门作为主要的区隔手法，私人空间的隐秘性得以保留，且灵活的区隔让空间尺度放大。没有视线的阻隔，在无形中将空间组合起来，因而有了自由而宽敞的居住环境，让留白成为住宅中最无价的奢华。

A clean background, without the presence of excessive signage, is the first impression this property gives to visitors. The designer ditches the excessiveness of having too much design, so that the characteristic of the space can return to its purest state. There are a lot of intricacy and sophistication in the design: it is evident by the fact that messy surface and complicated cut lines are nowhere to be seen. This kind of design technique helps to produce a sense of freedom in the living space.

The staircase acts as an intermediary point of the house. The living room and the dining room are located at the left-hand side and right-hand side from the entrance, respectively. Visual obstacles are not present in the design. Under the effect of natural lights, the cleanness and expansiveness of the space is emphasized. Contour line is a main element in most interior designs. However, the designer pays more attention to the overall appearance of the walls and thinks that the main focus should lay in having the right proportions in relation to the space available. With mathematical precision, the designer measures the length and width of each material used carefully. As for surface material such as stone, component made of iron, as well as imported porcelain tile; all brings enjoyment to the senses by the superior feel they provide. In this living space, outstanding quality and good taste are thoroughly displayed.

In terms of the spatial design, sliding doors are used as dividers. The privacy of the personal space is retained in this way, but at the same time allows the illusion of a larger space, with visual obstructions. With all these thoughtful designs, a free and spacious living environment is created. In this fashion, "leave it blank" has now become the most priceless luxury you can have in a house.

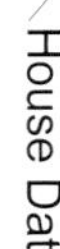

空间案名　许宅
设 计 者　王玉麟
参与设计　戴玉铃
设计公司　MPI设计公司
空间地点　新北市
空间性质　独栋别墅
空间面积　513 m^2
空间格局　玄关、客厅、餐厅、厨房、书房、主卧室、卧房×2、主卧更衣室、主卧浴室、浴室×3
主要建材　钢件、石材、进口石英砖、玻璃、木皮、铁件、硅酸钙板

Case Name / Hsu's House
Designer / Yu-Lin Wng
Co-designer / Yu-Ling Tai
Company / MPI design
Location / Taipei County
Nature of the apartment / Detached villa
Area / 513m^2
Outline / entrance, living room, dining room, kitchen, study room, master bedroom, bedroom x2, master dressing room, master bathroom, bathroom x3
Main building material / Steel, stone, imported porcelain tile, glass, wood, iron, calcium silicate board

01 02 03 04 05 06

01 _ 楼梯间　02 _ 起居室　03 _ 主卧室　04 _ 3F廊道
05 _ 主卧室　06 _ 厨房
01 _ stairway　02 _ living room　03 _ master bedroom
04 _ 3F corridor　05 _ master bedroom　06 _ kitchen

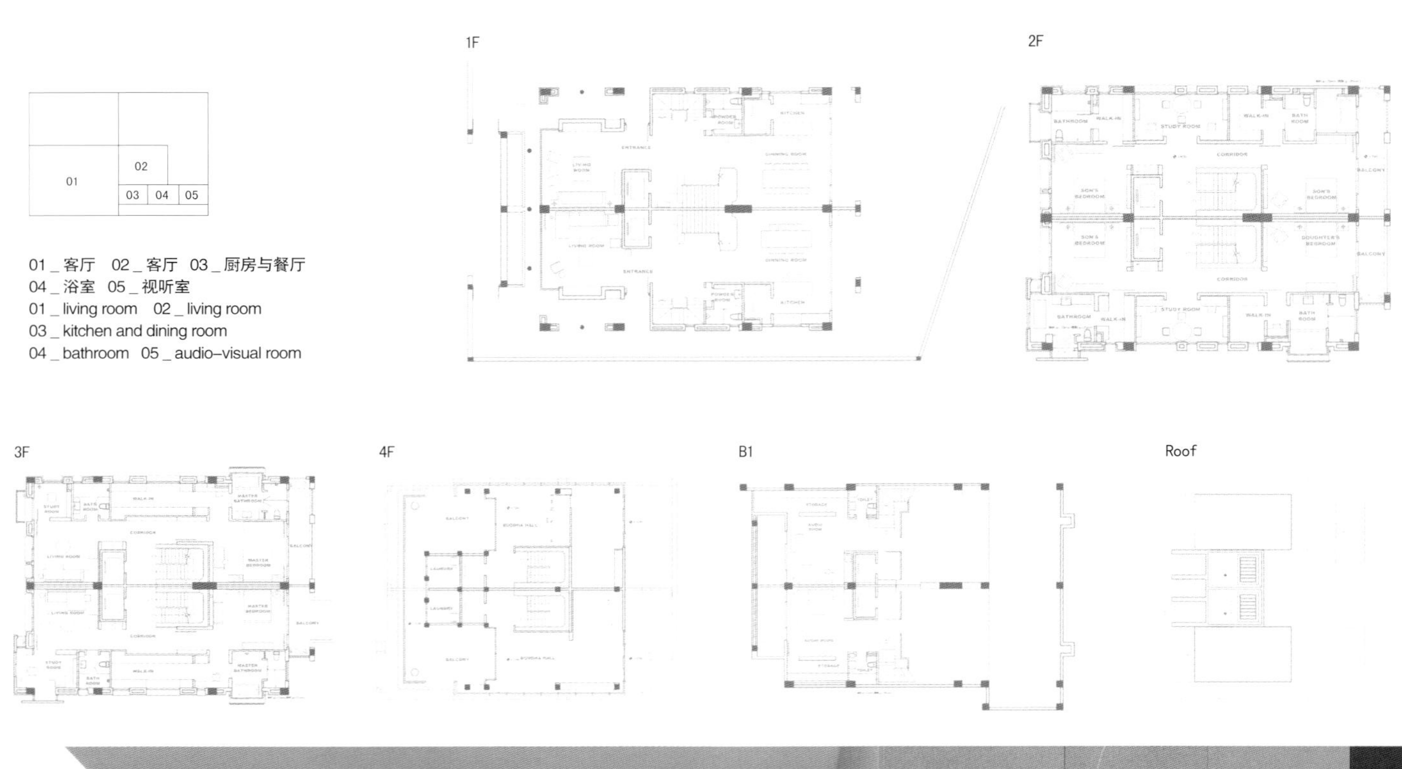
01
02
03
04
05
01 _ 客厅　02 _ 客厅　03 _ 厨房与餐厅
04 _ 浴室　05 _ 视听室
01 _ living room　02 _ living room
03 _ kitchen and dining room
04 _ bathroom　05 _ audio-visual room
1F
2F
3F
4F
B1
Roof

Designer Data

王玉麟
现　任／
MPI设计公司执行长
学　历／
法国巴黎艺术学院毕业
经　历／
MPI设计集团（美国纽约）
IFI国际设计师联盟会员
简　介／
2003～2005年连续获得三届日本JCD设计大奖，2007、2008年获选为TID台湾设计奖TID奖、公共空间类金奖以及临时建筑类 TID奖。现为台湾观光主管部门观光旅馆质量监评委员、中原大学室内设计系助理教授。

Yu-Lin Wang
Current Position
Executive, MPI-DESIGN
Education
Ecole des Beaux-Arts, France
Experiences
MPI -DESIGN GROUP NEW YORK USA
IFI international designer's union, affiliate
Brief bio
JCD Design Award (in Japan) for three consecutive years, from 2003～2005; TID Design Award (Taiwan), Gold Award for Public Space Design, TID Award: Provisional Building Award, for 2007 and 2008; currently takes the position as the Hotel Quality Assessment Committee Member for Taiwan as well as being the Assistant Professor of Chung Yuan Christian University.

释放尺度，创造自由宽敞的生活空间

Expand ! Create a spacious and free living space

本案位于台北市楼房密度低的住宅区，屋主期望在空间的功能性需求能被满足之余，还能拥有高质感与大尺度的特性。空间大尺度可借由视线延伸达成，再通过轴线的设定以及恰如其分的格局规划，表现出属于都会大宅的独特气势。

本案空间为两户合并，好客的屋主期待这个宽敞的空间有独立的宴会场所，因此在空间格局的配置上，分为公共与私人两个区块。宽敞的餐厅与主卧房自成一区，开放式的厨房可满足基本所需，让空间的呈现更臻完整。设计师通过动线的规划，将空间尺度释放，通过拉门作为空间区隔的手法带来视线的延伸，空间的宽敞度因而被显现。

在此空间中，设计的细节来自于屋主对生活态度的抽丝剥茧，风格内敛却又具有丰富的设计语汇。活动自如的空间创造自由的生活方式，也提供居住者更多元的选择。设计师通过空间的设计，带给居住者新的惊喜——去面对自己熟悉却又创新的家。

This apartment is located at a low-density residential area in Taipei. As the demand for space and functionality has already been satisfied, the owner wants something more in life: a high quality environment that is also large in scale. A sense of space is created through the extension of the visual horizon. Through the use of axis and appropriate layout planning, a unique character that is both urban and spacious is created in the apartment.

The property is actually two apartments renovated into one. The hospitable house owner wishes that there is an independent place for banqueting within this capacious space. For this reason, the structural layout of the space is divided into two zones, public and private, respectively. The spacious dining room and the master bedroom are contained their own zones. An open kitchen also meet their basic need, in the same time allows the "completeness" of the space. The designer uses special route planning in order to create a feeling of large space. An extension in the visual space is created by using sliding doors as room dividers. This gives an illusion of an even larger space.

Within this space, the inspiration for the design comes from a careful investigation of every details of the owner's attitude towards life. The style itself is quite low-key yet sophisticated in design. A space in which the owner of the house can move freely helps to create a carefree life. The designer brings many new surprises to the residents through space reconstruction. This incites familiarity as well as innovation in the mind of the house-owners when they look at their new home.

Photo courtesy_ MPI design Photographer_ Marc Gerritsen
图片提供_MPI design 摄影_Marc Gerritsen

dining room, kitchen 餐厅、厨房

House Data

空间案名　峰范
设 计 者　王玉麟
参与设计　陈亮瑄
设计公司　MPI设计公司
空间地点　台北市
空间性质　电梯大楼
空间面积　460 m²
空间格局　玄关、客厅、厨房×2、餐厅、主卧×2、起居室、客房×2、卫浴×4
主要建材　钢件、石材、玻璃、木皮、铁件、硅酸钙板、磁漆

Case Name / Feng Fan
Designer / Yu-Lin Wang
Co-designer / Liang-Hsuan Chen
Company / MPI design
Location / Taipei City
Nature of the apartment / Condominium
Area / 460m^2
Outline / entrance, living room, kitchen x2, dining room, double master bedroom, living room, two guest rooms, bathroom x4
Main building material / steel, stone, glass, wood, iron, calcium silicate board, enamel paint

01 _ 玄关　02 _ 客厅　03 _ 客厅与餐厅　04 _ 主卧室1
01 _ entrance　02 _ living room　03 _ living room and dining room
04 _ master bedroom 1

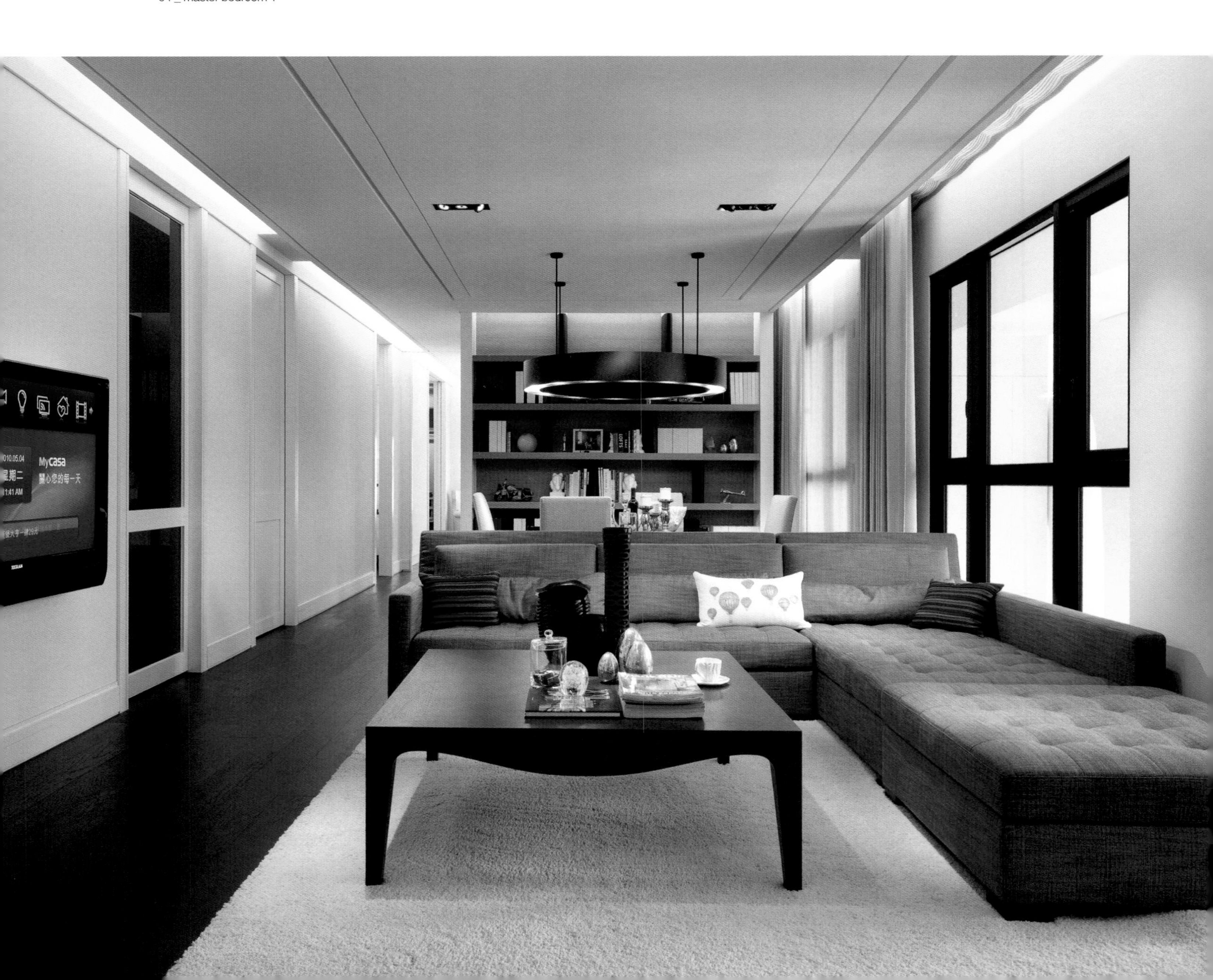

Designer Data

王玉麟
现　任/
MPI设计公司执行长
学　历/
法国巴黎艺术学院毕业
经　历/
MPI设计集团(美国纽约)
IFI国际设计师联盟会员
简　介/
2003~2005年连续获得三届日本JCD设计大奖，2007、2008年获选为TID台湾设计奖TID奖、公共空间类金奖以及临时建筑类TID奖。现为台湾观光主管部门观光旅馆质量监评委员、中原大学室内设计系助理教授。

Yu-Lin Wang
Current Position
Executive, MPI-DESIGN
Education
Ecole des Beaux-Arts, France
Experiences
MPI - DESIGN GROUP NEW YORK USA
IFI international designer's union, affiliate
Brief bio
JCD Design Award (in Japan) for three consecutive years, from 2003-2005; TID Design Award (Taiwan), Gold Award for Public Space Design, TID Award: Provisional Building Award, for 2007 and 2008; currently takes the position as Taiwan Hotel Quality Assessment Committee Member as well as being the Assistant Professor of Chung Yuan Christian University.

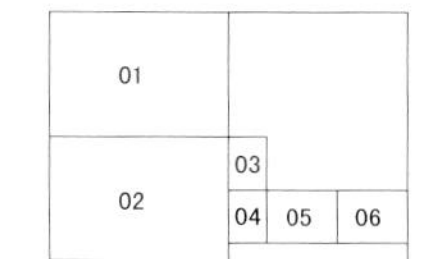

01 _ 主卧室2　02 _ 主卧室1　03 _ 主卧更衣室　04 _ 起居室　05 _ 主卧室1　06 _ 主卧室2
01 _ master bedroom 2　02 _ master bedroom 1　03 _ master dressing room
04 _ living room　05 _ master bedroom 1　06 _ master bedroom 2

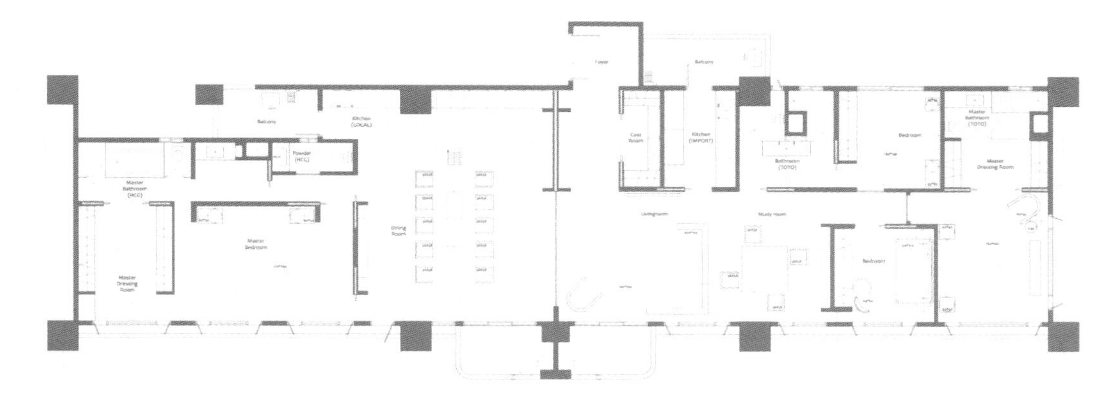

独立，拥有更多的奢华享受

Be independent, and embrace more luxury

outer appearance外观

living room 客厅

对于屋主而言，家不只是一个遮风避雨的地方，而是一个完全私有的生活场所。因此，唯有构筑出一个完美的空间，才能确保给家人一个最舒适的居住环境。依循这样的概念，这间别墅在功能与格局配置上都有着精确的规划，通过空间的定位，设计师替屋主打造出一个全然独立的豪华住宅。

考虑到大房子中除了屋主一家外，也有长辈同住，以及亲朋好友不定时来访，故在格局的设定上，屋主一家的生活重心设定于二楼，客厅与小吧台的配置，让屋主一家在二楼可拥有完整的生活空间。而客房与长辈房分别设定于一楼与三楼，除了可各自独立，电梯的设置也让长辈的出入更为便利。

一楼主要为公共空间，玄关通过展示柜以及屏风的设计，展现出宽敞豪华的气势。客厅主要作为接待访客之用，而开放式厨房与餐厅的设计，更展现出家的宽阔感。另外，男主人专用的书房采用玻璃隔断，除让公共空间更显得自由通透，同时也与庭院连接，将室内外的空间串联起来，展现出一般公寓所难以拥有的宽敞动线。

For the house-owners of this design case, home is not just a place to shelter, but a completely private living space. For this reason, only by building a complete building, can ensure that your family a most comfortable living environment. Based on this concept, equal weights are placed on the functionality and the spatial layout. Through the partitioning of the space, the designer has created an independent and luxurious home.

From the point of view of space allocation, the designer has considered the fact that in addition to the nuclear family of the house-owners, their elderly parents also live with them. Furthermore, relatives and friends who visit the house from time to time also require accommodation. Therefore, in turns of the spatial layout, the main living quarters lay on the second floor. The building of a living room and a small bar at the second floor allows the family to enjoy a comprehensive living space. As for the guest's room and the elder's room are located on the first floor and the third floor, respectively. Not only are these rooms independent from each other, the building of the elevator also allows easier access for elderly parents.

Floor one is used as the public space; the use of the display cabinet and folding screen at the entrance also makes the design look more spacious and luxurious. The main function of the lounge on the first floor is primarily used for the reception of visitors. Furthermore, the expansiveness of the home is also exhibited by the open-style kitchen and dining room. Moreover, glass panel is used as the divider for the host's private study room. This in turn creates an open and more transparent public space. Finally, the room is also joined with the outdoor garden. This helps to link the indoor and outdoor environments together, fully exhibiting a spacious route design that can hardly be done at an ordinary apartment.

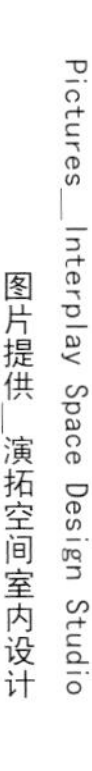

House Data

空间案名 夏悠宫王宅
设 计 者 张德良、殷崇渊
参与设计 演拓空间室内设计
空间地点 台北市
空间性质 别墅
空间面积 434 ㎡
空间格局 1楼／客厅、厨房、餐厅、书房、客房、玄关、储藏室；2楼／起居室、小孩房×2、主卧室、主卧卫浴、小孩房共享浴室；3楼／神明厅、起居室、长亲房、洗衣房
主要建材 抛光瓷砖、大理石、美耐板、茶色镜、木皮、金属箔、耐磨地板、实木集成材

Case Name / Chen's House
Designers / Te-Liang Chang, Chung-Yuen Yin
Company / Interplay Space Design Studio
Location / Taipei County
Nature of the apartment / villa
Area / 434m²
Outline / 1F louge, kitchen, dining room, study room, guest room, entrance, storage room; 2F living room, kids' room x2, master bedroom, master bathroom, kids' shared bathroom; 3F Tao Religious Altar, living room, elder parents' room, laundry room
Main building materials / polished tiles, marble, melamine plates, tea mirror, wood, gold leaf, hard-wearing flooring, laminated wood

01 _ 玄关 02 _ 2楼客厅 03 _ 客厅 04 _ 餐厅与书房 05 _ 餐厅与厨房
01 _ entrance 02 _ 2nd floor lounge 03 _ living room 04 _ dining room and study room 05 _ dining room and kitchen

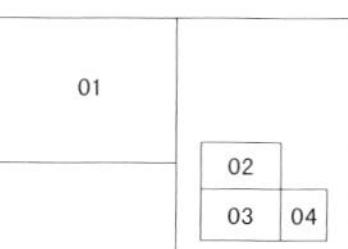

01＿餐厅　02＿主卧室　03＿主卧浴室　04＿客浴
01＿dining room　02＿master bedroom　03＿master bathroom　04＿guest bathroom

Designer Data

张德良、殷崇渊
现　任/
演拓空间室内设计 总监
经　历/
台北市室内设计装修商业同业公会会员
1999 创立演拓空间室内设计
简　介/
坚持好的设计应是体贴使用的人。设计的目的在于使生活简单，而不是让居住者花过多的时间整理照顾，不应该让不合宜的设计造成居住者的负担。

Te-Liang Chang, Chung-Yuen Yin
Current Position
Director, Interplay Space Design Studio
Experiences
Member of Taipei Association of Interior Designers
1999- Established Interplay Space Design Studio
Brief bio
A good designer should be considerate towards his/her client. The design aims to make life simpler, rather than letting the residents spending too much time taking care of their property; and should not cause burden for the inhabitants with inappropriate designs.

好莱坞风格的观景大宅

A hollywood style observatory residence

living room 客厅

连同地下室共4层楼近412m²的别墅，设计师将空间重新配置，地下一楼规划为厨房、餐厅及起居室；一楼除了客厅外，还规划了阅读区、休憩区；二楼为女儿房及起居室；三楼则完全为屋主所有，主卧＋主卧专属更衣室＋主卧浴室。超越六星级饭店的配备，搭配窗外自然美景，家就成为最佳的减压空间，在家就跟在度假一样。垂直空间以电梯串联，让生活更为便利。

顺应客厅6m挑高，电视主墙以雪白银狐大理石打造，展现出大宅的气势。大理石是大宅必备的材质，设计师特别挑选雪白银狐大理石装饰壁炉，搭配镶着雨花白苏菲亚大理石的地板，更突显空间气度。

客厅角落规划壁炉，满足功能及风格的需求。客厅天花板造型结合空调出风口设计，并以线板装饰，通过材质及设计语汇展现风格的细致及气质。

The villa, combined with the basement, has a total of four stories and 412 m^2 of space. The designers redesigned the spatial compartments, assigning basement 1 as the kitchen, dining room and living room; the 1st floor will include a study and a resting room in addition to the living room; the 2nd floor would be the lady's dressing chambers and living room; the 3rd floor would be the personal space of the resident, with a main bedroom, a main dressing room and a main bathroom. The quality of the facilities exceeds that of a 6 star hotel combining a priceless and beauteous view of the surrounding landscape. The house is therefore transformed into a holiday resort, the perfect place for rest and cooling off from stress. An elevator connects the vertically stacked living compartments for an easier life.

A fitting material for the six meter tall living room of snow fox marble is chosen for the television wall, giving the room the spacious openness of a large villa. Marble is considered the primary choice for the interior design of villas; hence a set of carefully chosen snow fox marble lines the fireplace combined with white splattered Sofia marble flooring is used to enhance the sense of volume.

The fireplace designed into the corner of the living room both serves a function and a style; the ceiling is decorated with lining boards which integrates the A/C openings into a part of ceiling designs, expressing a stylish delicacy and elegance with both material choice and style.

Photo provided_IS International Design
图片提供_IS国际设计

living room 客厅

House Data

设 计 者	陈嘉鸿
设计公司	IS国际设计
空间地点	新北市
空间性质	别墅
空间面积	412 m^2
空间格局	地下1楼：电梯间、厨房、餐厅、起居室、客房、佣人房、客用浴室；1楼：电梯间、玄关、衣帽间、客厅、起居室、客用浴室；2楼：电梯间、女儿房、浴室、书房；3楼：电梯间、主卧、主卧更衣室、主卧浴室
主要建材	热带雨林大理石、金镶玉大理石、夜光石大理石、琥珀蓝宝大理石、咖啡绒大理石、雪白银狐大理石、雨花白大理石、苏菲亚大理石、深金峰及浅金峰大理石、黑檀木木皮、高级喷漆、灰色镜、茶色镜、进口皮革、进口瓷砖

Designer / Chia-Hung Chen
Designer / Chia-Hung Chen
Company / IS International Design
Location / New Taipei City
Type of space / Villa
Area / 412 m^2
Layout / B1: Elevator room, kitchen, dining room, living room, guest room, Maid's Chambers;1F: Elevator room, front porch, closet, living room, lounge, guest bathroom;2F: Elevator room, lady's dressing chambers, bathroom, study;3F: Elevator room, master bedroom, master changing room, master bathroom
Main materials / Rainforest marble, gold-in-jade marble, luminous marble, amber-sapphire marble, coffee velvet marble, snow fox marble, splattered white marble, Sofia marble, deep gold-peaks and light gold-peaks marble, ebony bark, high grade spray paint, gray-colored mirror, tea-colored mirror, imported leather, Italian tiles

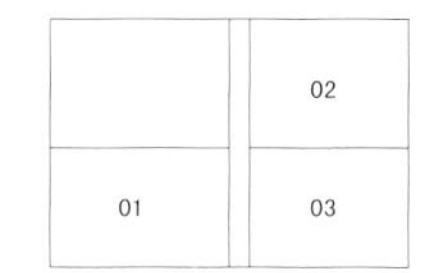

01 _ 客厅 02 _ 客厅 03 _ 客厅角落
01 _ living room 02 _ living room
03 _ living room corner

01	03	
02		06
04	05	07

01 _ 梯间　02 _ 玄关　03 _ 梯间　04 _ 餐厅　05 _ 主卧　06 _ 梯间　07 _ 浴室
01 _ intermediate landing　02 _ front porch　03 _ intermediate landing　04 _ dining room
05 _ master bedroom　09 _ intermediate landing　07 _ bathroom

Designer Data

陈嘉鸿
现　任／
IS国际设计　主持设计师
简　介／
凭借天生的空间感和独特的美学品味，投入室内设计的领域，擅于运用垂直及水平线条来捕捉空间向度，并常以隐藏式的收纳方法及家具的搭配，打造出独特的居家风格。
著作／
《室内设计师陈嘉鸿的隐藏学》

Designer Data
Chia-Hung Chen
Current employment
Lead Designer, IS International Design
Resume
Chen combines an innate acute spatial sense with a unique set of aesthetics, and is fully devoted to the field of interior design. His precise handling of scale and skilled use of vertical and horizontal lines allows an accurate depiction of spatial direction. The hidden style of storage methods combined with the furniture creates a unique residential style.
Publications
Hiding Methods of Interior Designer Chia-Hung Chen, published by MyHomeLife

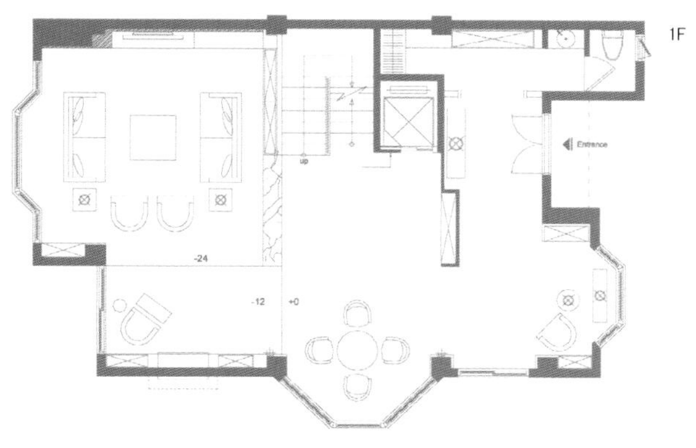

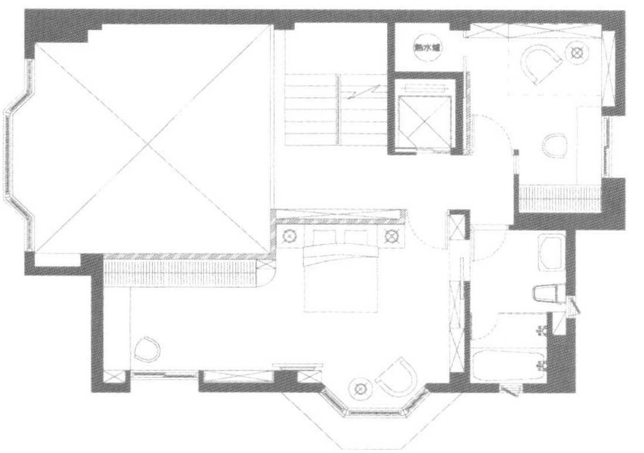

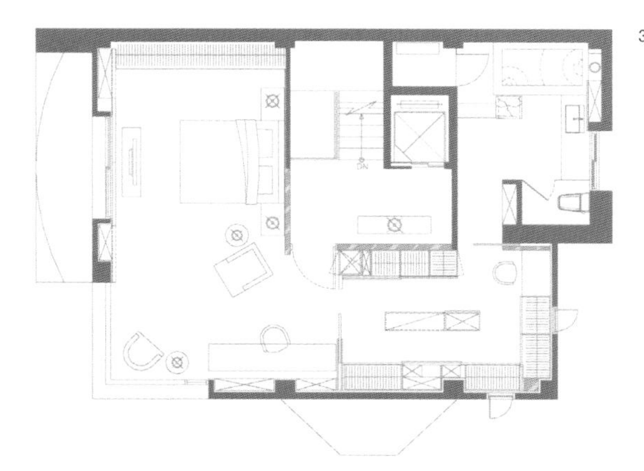

living room 客厅

开放古典，创造旷达的生活空间

Create an expansive and classic living space

living room 客厅

屋主偏爱美式乡村风格，期望通过设计师的帮助，将该独栋别墅打造成原味浓烈的美式居家。考虑到建筑本身的结构，设计师通过天窗的设计引进自然光线，让大宅不只具有宽敞的尺度，也因采光良好而展现出清新的质感。

为了让空间风格更贴近美式乡村风，全室家具均由加拿大进口，通过实木等材质的搭配运用，营造出舒适悠闲的气氛。由于原有的楼梯过于死板，设计师刻意重新以旋梯形式设计，让空间更具优雅质感。客厅运用古典的对称元素，展现方正宽广的格局。而厨房与餐厅之间，通过拉门区隔，并以开放式手法让空间放大，流畅的动线创造出自由的空间，更加彰显大宅的宽阔气势。

本案善用建筑格局将自然采光优势放大，借助动线的串联与开放式的设计，释放空间尺度，替屋主打造出理想的居住环境，也彰显大宅之所以大的价值。

Due to the house owner's strong liking for "American country style", he wants a designer who can transform his new home into an American-styled mansion. After a careful analysis of the structure of the building, the designer decides to use skylights to bring natural light into the house. This not only gives an illusion of a larger space in an already large house, but also provides the surroundings a refreshed and pleasant vibe.

In order to fully incorporate the "American country style" into the house design, all the furniture are specially imported from Canada. Materials such as line board and hardwood are used interchangeably to create a comfortable and leisurely environment. The designer thinks that the original design of the staircase looks too rigid; for this reason, he reconstructed it into spiral design, which in turn makes the whole space feel more elegant. The living room is designed in a symmetrical way as can be found in traditional, classical design, giving the space a broader, more squaresque feel. As for the design of kitchen and dining room, sliding doors are used to make the space look more open but at the same time further visually magnify the room. Furthermore, the employment of smooth routes also helps to create a freer space, further make the house look grand and imposing.

In summary, the main design concept of this house is to take advantage of the original structural layout to provide an ample supply of natural lighting. Moreover, by using an open-design as well as interconnecting the routes, an expansive space is created. In any case, the designer has built an ideal living environment for the house owner, as well as defining the true meaning of a mansion.

Pictures_Dehan International Interior Planning Ltd.
图片提供_大涵国际室内企划有限公司

living room 客厅

House Data

空间案名　张公馆
设 计 者　赵东洲
参与设计　赵晏良、林纯萱
设计公司　大涵国际室内企划有限公司
空间地点　别墅
空间性质　电梯大楼
空间面积　396 m^2
空间格局　五房三厅四卫
主要建材　石英砖、复古砖、大理石、柚木、南方松木、耐火壁炉砖、水晶、烤漆处理实木

Case Name / Zhang Residence
Designer / Tung-Chou Chao
Co-Designer / Yen-Liang Chao, Chun-Hsuang Lin
Company / Dehan International Interior Planning Ltd.
Location / Taoyuan County
Nature of the apartment / villa
Area / 396m^2
Outline / bedrooms x5,living rooms x3, bathrooms x4

01 _ 餐厅与厨房　02 _ 餐厅　03 _ 餐厅　04 _ 书房　05 _ 书房
01 _ dining room and kitchen　02 _ dining room　03 _ dining room
04 _ study room　05 _ study room

Designer Data

赵东洲

现　任／

大涵国际室内企划有限公司设计总监

鸿室空间设计有限公司设计总监

台湾室内装修专业技术人员学会副理事长

台湾室内设计装修商业同业公会全国联合会理事

台北市室内设计装修商业同业公会常务理事

中原大学、中原大学研究所专家级助理教授

台湾建筑物室内设计技能检定命题委员

媒体专栏作家

学　历／

中原大学室内设计系硕士

经　历／

大仁室内计划公司项目经理

台湾装修技术人员协会秘书长

台湾室内设计协会理事

台湾中国科技大学室内设计系兼职讲师

华夏技术学院室内设计系兼职讲师

Tung-Chou Chao

Current Position

●Designer Executive of Dehan International Interior Planning Ltd. Company

●Designer Executive of Hong-Shih Interior Design Ltd. Company

●Vice Board Chairman of Chinese Association of Interior Design

●Board Chairman of Taiwan Association of Interior Design

●Managing Director of Taipei Association of Interior Designers

●Assistant Professor of Chung Yuan Christian University and Research Institute of Chung Yuan Christian University

●Examiner, Interior Design Skill Assessment, Council of Labor Affairs of Executive Yuan

●Columnist

Education

●Master of Interior Design, Chung Yuan Christian University

Experiences

●Project Manager of DAZA Interior Planning Co., Ltd.

●Secretary of Chinese Association of Interior Designers

●Director of Chinese Society of Interior Design

●Adjunct Instructor of Interior Design, Chinese University of Technology

●Adjunct Instructor of Interior Design Hwa Hsia Institute of Technology

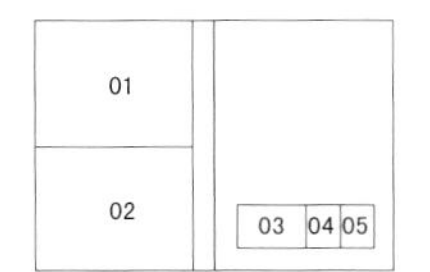

01＿厨房　02＿起居室　03＿主卧室　04＿阁楼　05＿浴室

01＿kitchen　02＿living room　03＿master bedroom

04＿attic　05＿bathroom

精致奢华，打造质感满分的大宅

Refined and luxurious：create a perfect home

dining room 餐厅

dining room 餐厅

居家空间的奢华，未必来自于高级建材的使用，有时候其珍贵之处，正在于能将旧有的记忆留存，并与新的设计互相融合。屋主为了能给女主人与小孩一个更为完整的生活空间，因此将住家重新翻修。设计师通过色调、图案的运用，加上将原有的柚木地板重新处理，为屋主一家打造出温暖而精致的生活场所。

在本案中，可以看到设计师精心运用各种素材，铺陈出大宅的独特样貌。玄关以精致的大理石拼花以及弧形天花板垂吊的水晶灯，营造出丰富的视觉效果。而将屋主收藏的画作摆放在端景墙面上，除能以艺术品提升空间质感，也通过屋主的收藏展示，让家的意义更为丰富。

大宅中的精致元素，在本案中也有相当多元的呈现。如将皮革、施华洛世奇水晶等特殊材质应用在房间的主墙上，木作上使用激光切割的图案，或以特殊的珍珠白色喷漆处理，线板上运用手工贴饰金箔，厨房收纳柜的拉门采用雕刻玻璃，这些设计让空间从奢华的风格中延展出更为优雅的姿态，也替大宅作了更不一般的诠释。

The luxuriousness of the living space doesn' t necessarily come from the use of building materials; what is the most precious is to have your old memories intact, and incorporates these memories into the new design. Since the house owner wants a more all-inclusive living space for his wife and children, he decided to let his house renovated. Through a careful selection for the right colors/tones and patterns, in addition to the retreatment of the existing teak wood flooring, a warm and elegant home is created.

In this property, various materials are carefully selected in order to build a unique appearance. At the entrance, a rich visual effect is created via the use of marbles and a chandelier. The owner' s collection of paintings is also mounted on the side wall. Not only does it make the whole place artistic, but also makes the owner feeling more at home.

Fine elements found in a well-designed property can also be found here. The designer uses unusual materials such as leather and Swarovski crystals on the main wall of the room. Furthermore, his decision to use laser to engrave patterns onto woodworks; to use a special pearl-colored paint for spray-painting; to apply hand-crafted gold leaf onto the line boards, and to use sliding door made of carved glass on the kitchen cabinet, which fully exhibits the beauty of transparency, all helps to upgrade the home from being simply extravagant to a home where luxury and elegance are met. In summary, this design really redefines the meaning of a home.

图片提供_权释国际空间设计 Pictures_Allness Design

House Data

空间案名　御之苑
设 计 者　洪韡华
设计公司　权释国际空间设计
空间地点　台北市
空间性质　电梯大楼
空间面积　333 m²
空间格局　外玄关、内玄关、客餐厅、厨房、男主人房、女主人房、起居室、3间儿童房、浴室×5、儿童阅读游戏区、佣人房
主要建材　纯手工香槟金属箔、进口壁布壁纸、马鞍皮、激光切割图案造型、古典线板、铁件、喷漆、气垫地板、柚木地板、雪白玉、橄榄黑、流星雨、结晶化玻璃、施华洛施奇水晶灯、多乐士竹炭漆、欧诗木涂料、烤漆柚木、特殊图腾处理茶色镜、超白烤漆玻璃、超白玻璃特殊色烤漆、钻雕玻璃

Case Name / Yu Tze Fan
Designer / Wei-Hua Hung
Company / Allness Design
Location / Taipei City
Nature of the apartment / Condominum
Area / 333m²
Outline / outer entrance, inner entrance, dining room, kitchen, host's room, hostess' room, living room, children's rooms x3, bathrooms x5, Children's reading / playing area, servant's room
Main building material / hand-crafted champagne gold leaf, imported wallpaper, saddle leather, laser-cut totem design, classical line board, iron, spray paint, air cushion flooring, teak wood floors, white jade, "olive black", "meteor rain", crystal glass, Swarovski crystal, ICI charcoal paint, OSMO paint, teak steel grill, special totem processing for the opaque mirror, ultra-white painted glass, ultra clear glass special colored enamel paint, diamond-cut glass

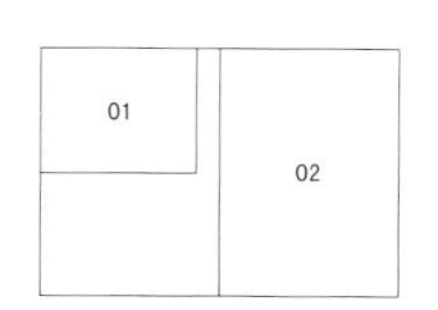

01 _ 客厅　02 _ 玄关
01 _ living room　02 _ entrance

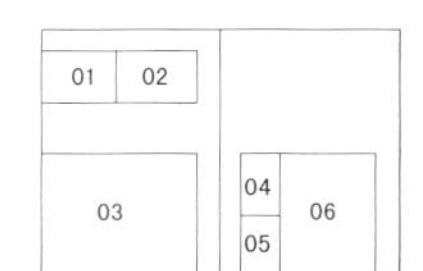

01＿客厅　02＿玄关　03＿女主人房　04＿浴室　05＿女主人浴室　06＿男主人房
01＿living room　02＿entrance　03＿hostess' room　04＿bathroom
05＿ hostess' bathroom　06＿ host's room

Designer Data

洪韡华
现　任/
权释国际空间设计 总监
简　介/
擅长依循屋主不同性格打造出深具东方诗意居家情境，且也能设计出最耐人寻味且百看不厌的"新大宅"样貌。著有《我家就是五星级饭店—— 提升居家质感设计200》

Wei–Hua Hung
Current Position
Director, Allness Design
Brief bio
My specialty are to create a personalized living environment according to the personality of my client, in oriental style. I am also good at designing homes that are contemporary in style, and which you would never get bored by looking at it. I have written a book called "My home is a five–star hotel–enhance your home designing 200".

打造高质感的女性私宅

Building a high quality private apartment for an independent woman

living room, dining room and kitchen 客厅、餐厅与厨房

图片提供 | 权释国际空间设计

Pictures | Allness Design

为事业有成的女强人打造专属的住宅，需要多一份细腻。设计师在本案设计之初，以"珠宝盒"的概念，作为替双拼豪宅量身设计的主轴，整体空间呈现法国凡尔赛宫的华丽与优雅，打造出专属女主人的私人宫殿。

对于女主人而言，家是一个可以放松休憩的地方，也是可以让她将珍爱的收藏品尽情展示的场所。因此在公共空间的区块，设计师将女主人珍藏的画作与家具摆放在空间一角，并赋予宽敞的展示舞台。而通过银狐大理石的使用，加上印度黑大理石滚边，将整个空间的气势撑起。浅色系的运用，及线板的搭配，衬托出古典风格空间的高雅。通过将壁炉与电视作为中轴线，客厅与餐厅空间因而得以区隔，也呈现出平衡的美感。

房子为两个大宅打通的格局，女主人使用的区域，宛如在家另辟的贵宾室。宽敞的主卧内有如精品柜的更衣室以及浴室，除了有干湿分离的浴室之外，还另设有水疗室，让女主人回到家也能享受超五星级饭店的生活质量。

When designing an apartment for a strong woman with a successful career, you should pay an extra attention to the details. The designer uses the concept of a "jewelry box" when designing this apartment. The design is tailor-made to suit the hostess so that the whole design rivals that of the Versailles of France in terms of its extravagance and elegance, making this apartment, which is actually two apartments renovated into one, an exclusive and luxurious "private palace" for the hostess.

For this woman, home is a place where you can relax, as well as being the place where you can exhibit all your treasured collections. Taking this into account, the designer provides a large space where the hostess can display her cherished paintings and furniture. By using the Ariston marble as the base, then applying the Indian Black Marble to surround it, the whole place looks imposing, and in the same time, sophisticated and graceful, in a classical sense, as light-colored tone and line boards are also used to tone down hard lines of the marbles. The fireplace and television are used as a mid-point in which the living room and dining room are partitioned. The beauty of balance is fully exhibited here.

The apartment itself is actually two apartments renovated into one. The hostess' room looks as though there is a VIP room within it. Her spacious bedroom contains a dressing room that looks like a fine cabinet. As for her bathroom, not only is it partitioned into "wet and dry zones", there is also a separate spa room next to it, so the hostess can enjoy a lifestyle superior than a living in a five-star hotel.

House Data

空间案名　尚海陈宅
设 计 者　洪韡华
设计公司　权释国际空间设计
空间地点　新北市
空间性质　双拼住宅
空间面积　330 ㎡
空间格局　客厅、吧台、餐厅、玄关、厨房、水疗区、水疗区卫浴空间、主卧、主卧更衣室、小孩房2间、主浴室、客浴
主要建材　银狐大理石、黑云石、仿珍珠鳄鱼皮、进口窗帘、进口家具、大理石、壁纸、奥罗拉大理石

Case Name / Shang Hai Chen's house
Designer / Wei–Hua Hung
Company / Allness Design
Location / Taipei County
Nature of the apartment / two apartments renovated into one
Area / 330m²
Outline / living room, bar, dining room, entrance, kitchen, SPA area, SPA area en suite, master bedroom, master dressing room, children room x2, master bathroom, guest's bathroom
Main building material / Ariston marble, dark stone, pearl crocodile skin, imported curtains, imported furniture, marble in mosaic patterning,wallpaper, Aurora Marble

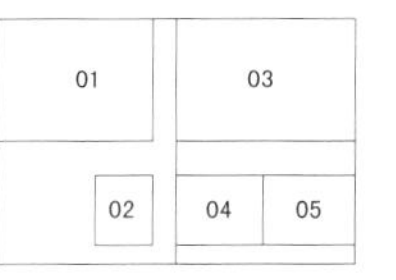

01 _ 客厅与餐厅　02 _ 餐厅　03 _ 客厅
04 _ 客厅　05 _ 厨房
01 _ living room and dining room
02 _ dining room　03 _ living room
04 _ living room　05 _ kitchen

01	02	03
04	05 / 06	07

01 _ 主卧室　02 _ 主卧浴室　03 _ 主卧更衣室
04 _ 主卧室　05 _ SPA室　06 _ 主卧浴室　07 _ 男孩房
01 _ master bedroom　02 _ master bathroom　03 _ master dressing room
04 _ master bedroom　05 _ SPA Room　06 _ master bathroom　07 _ boy's room

Designer Data

洪韡华

现　任／

权释国际空间设计 总监

简　介／

擅长依循屋主不同性格打造出深具东方诗意居家情境，且也能设计出最耐人寻味且百看不厌的“新大宅”样貌。著有《我家就是五星级饭店——提升居家质感设计200》

Wei-Hua Hung

Current Position

Director, Allness Design

Brief

My specialty are to create a personalized living environment according to the personality of my client, in oriental style. I am also good at designing homes that are contemporary in style, and which you would never get bored by looking at it. I have written a book called "My home is a five-star hotel-enhance your home designing 200".

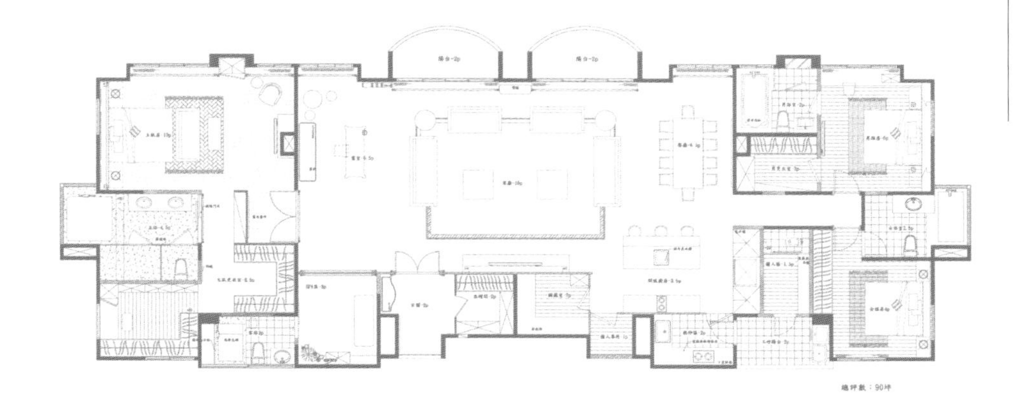

living room and dining room 客厅与餐厅

让家充满愉悦的氛围

Fill your home with pleasant atmosphere

试着去想象，一对年轻夫妻居住的别墅，如何能融合他们的生活背景与空间条件，营造出丰富的故事性？设计师充分利用建筑基地的优势，通过采光条件的扩展，营造出舒适的居住环境，让家充满爱和愉悦的氛围。

别墅从地下停车场到顶楼共七层，依照居住者的使用习惯，每一层都有不同的设计与视觉呈现。客厅与餐厅结合优越的采光条件，将空间格局做最恰当的定位，通透性与光线营造出宽敞舒适的空间感，餐厅与楼梯间的开窗，为起居空间绘出一道风景，楼梯与餐厅间激光切割图案的隔屏，更强调出了私宅的独特性。

男女主人的书房，细腻地融合使用者的个性以及曾经使用的旧家具，随兴的表现凸显出个人品位，也以混搭的手法揉合出多元的个性。女主人书房则与起居室相连，通过拉门的界定，也能让空间作为临时客房，活化了空间的功能。主卧室也因拥有独立的空间而有更自由的展现，放大的门框打破空间界限，将舒适的内涵表露无遗。

Try to imagine a villa owned by a young couple; how does the designer create a sense of architectural richness based on their background as well as the spaces available? By fully taking advantage of the superiority of the structural base, the designer helps to increase the effect of the lighting condition as well as offering a comfortable living environment, allowing the home to be surrounded by the atmosphere of love and joy.

The villa contains seven stories all the way from the underground garage to the top floor. According to the need of the inhabitant each floor has different themes and visual presentations. The living room and the dining room both contain superior lighting conditions, which is a bonus for the spatial arrangement of the design. The penetrative lights help to create a sense of expansive space. A window positioned in between the dining room and the staircase allows the inhabitants a clear view to the scenery outside. A tailored-made laser-cut pattern panel is used as a divider in between the dining room and the staircase, further emphasizing the uniqueness of the villa.

In the design of the host's study room, the element of the user's personality and the old furniture are delicately fused together, fully exhibiting his individual taste. Furthermore, mix and match approach is also employed to make the design more vivid. Moreover, the hostess' study room is connected to the lounge, which, through the positioning of sliding doors, can also be used as a temporary guest's room, further broadening its functionality. As for the master's bedroom, it is assigned to an independent space of its own, allowing a freer display. Finally, the enlarged door frame breaks down the boundaries of space, fully displaying the coziness of the house.

living room 客厅

Pictures| Simple Interior Design
图片提供＿尚展设计

01＿女主人书房 02＿起居室 03＿客厅 04＿厨房与餐厅 05＿餐厅
01＿ hostess' study room 02＿ living room 03＿living room
04＿kitchen and dining room 05＿ dining room

空间案名 文山名人刘宅
设 计 者 吴启民
设计公司 尚展设计
空间地点 台北市
空间性质 别墅
空间面积 330 m^2
空间格局 玄关、客厅、厨房、餐厅、女主人书房、起居室、男主人书房、主卧室、主卧浴室、主卧更衣室、客浴
主要建材 大理石、文化石、进口壁纸、柚木地板

Case Name / Wenshan-mingren Liu's Property
Designer / Chi-Min Wu
Company / Simple Interior Design
Location / Taipei City
Nature of the apartment / villa
Area / 330 m^2
Outline / entrance, lounge, kitchen, dining room, hostess' study room, living room, host's study room, master bedroom, master bathroom, master dressing room, guest's bathroom
Main building material / marble, great stone, imported wallpaper, teak-wood floors

1F

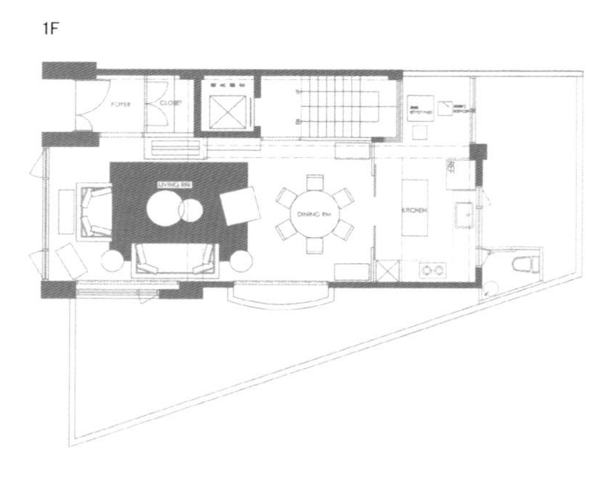

2F

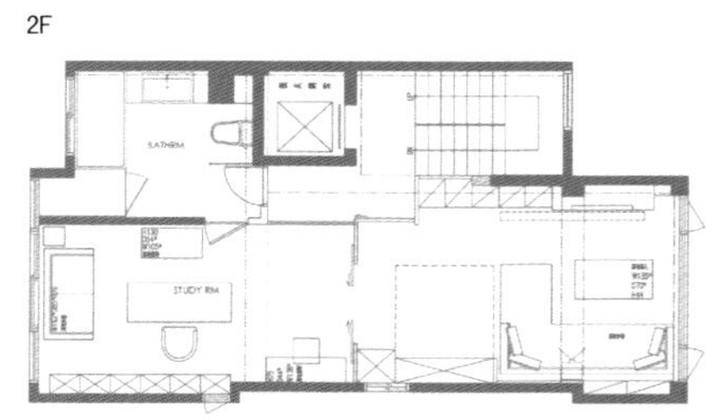

3F

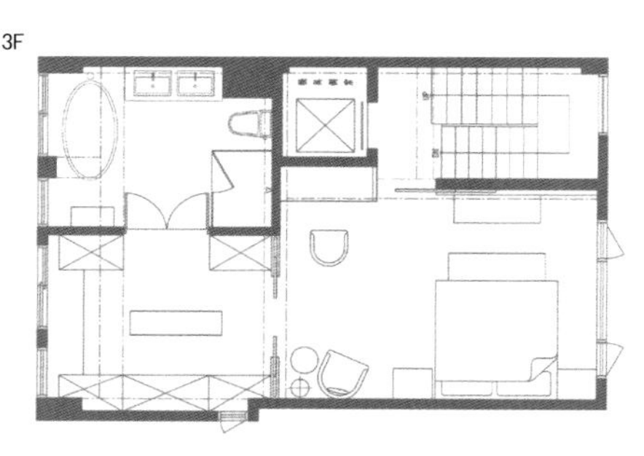

4F

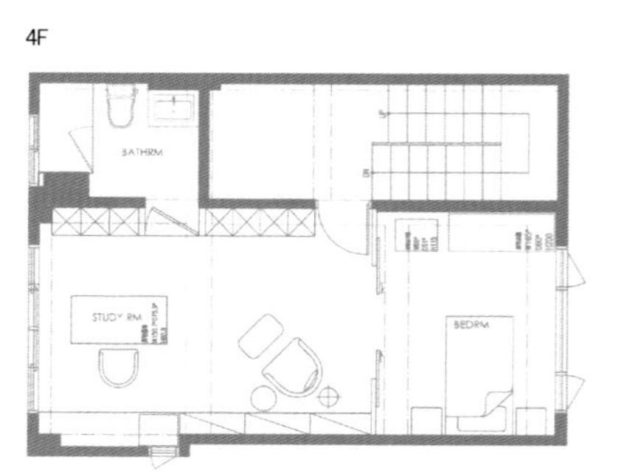

5F

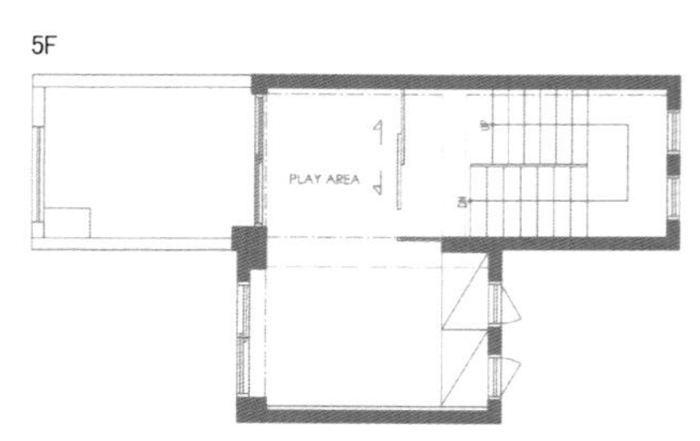

6F

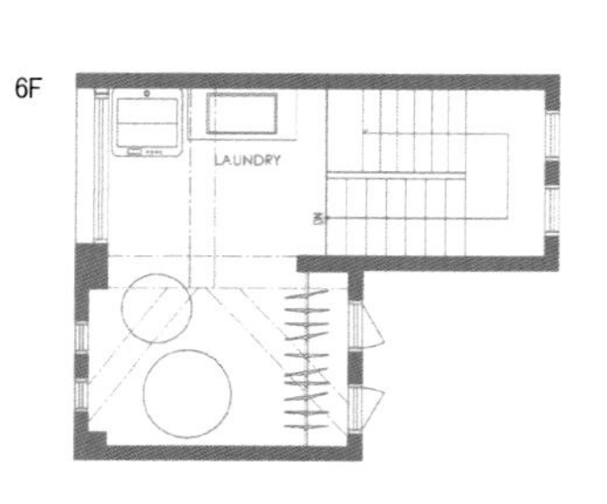

B1

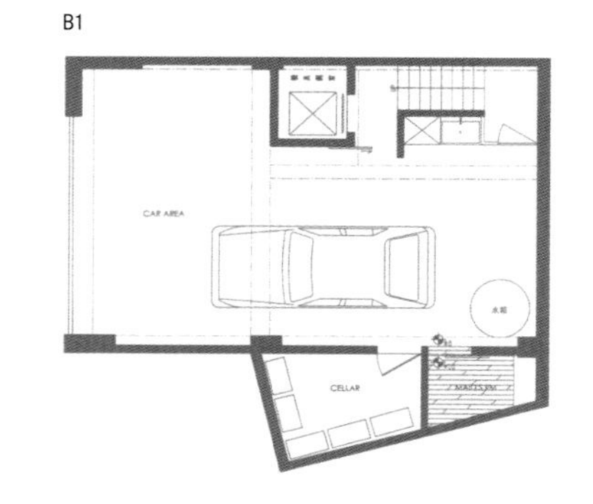

Designer Data

吴启民
现 任/
尚展设计 设计总监
学 历/
加拿大英属哥伦比亚大学（UBC）建筑与都市计划学系毕业
经 历/
汇侨设计、展亦设计 设计总监
2008年荣获AID Award 台湾室内设计大奖
2009年AID Award 台湾室内设计大奖入围
简 介/
始终坚持要利用房子本身的特性规划，顺着房子的优点来设计。有形的功能，回归自然变得纯粹。无形的价值，就不会被时间的潮流所淹没。奢华最极致的表现是融合自然与科技。着墨于每个微小的细节，让居住者的身心都能获得无比的满足与自在，这正是当前品味之士所追求的。

Chi-Min Wu
Current Position
Chief Design Officer of Simple Interior Design
Education
Bachelor of Architecture and Landscape Architecture, The University of British Columbia
Experiences
Chief Design Officer of Rich Honour Design Group and United Design Associate,
2008 AID Award
2009 Nominated AID Award
Brief bio
When designing for an apartment, I always strive to take the advantage of the original characteristic of the house. I insist to keep the good points intact. Functionality with forms allows the nature to become purer, so that values without forms would not disappear amongst the tide of time. The ultimate luxury is to combine technology and nature. I mind every single detail, so that both the mind and body of my clients can be satisfied and contended, the two values in which people with good taste would pursue.

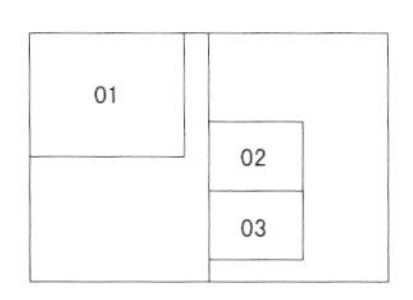

01 _ 主卧更衣室和主卧浴室 02 _ 男主人书房
03 _ 男主人书房
01 _ master bathroom and master dressing room
02 _ host's study room 03 _ host's study room

安定，让家大而温暖

Calmness makes your home big and warm

living room 起居室

Pictures_Hua-Shan Hsu Architectural Design Firm
图片提供_许华山建筑师事务所

本案的基地呈现不规则状，设计师善用原本的空间结构，将看似劣势的格局通过设计手法扭转。光线的导引以及线条的穿引，为这栋大宅营造出多层次的表情，而动线的设计，则带来如“刘姥姥进大观园”般的惊喜与独特性。

原有的基地结构因建筑的原因，较为琐碎且有死角，设计师通过光线的引导，让不同面向的光线都可进入室内，并且借此引导动线。如玄关入口处，透光不透明的屏风，就可将访客引入客厅。而私人空间由于拉门、玻璃的使用，也将光线引至室内，光线的流动，带来自由的韵律。即便空间随着基地结构而分成不同区块，但仍因光线而带来一种宽敞自由的印象。

在空间中，温暖而舒适的感受来自于材质的巧妙运用。设计师使用了老桧木制作的收纳柜、修理过的旧沙发，以及老柚木等材质，通过对视觉与嗅觉的亲和，营造出一种舒缓的气氛。而对线条的考究与整合，让天、地、壁产生对话，流畅的韵律也因此展开。光的穿针引线与材质间的对话，营造出一种极为安定的氛围，让舒畅进入居住者的心中。

The foundation of this design is irregular in shape. The designer effectively analyses the original structure, and reverse the seemingly inferior layout into something better. Natural lights are introduced to the apartment, which helps to make this it more multifaceted. Furthermore, the use of route lines also helps to bring a sense of excitement and uniqueness into the apartment.

The original base of the building contains many dead ends due to poor architectural design. For this reason, lights from different sources are guided into the room, which also helps to liven up the routes. Furthermore, the use of opaque yet light-transient screen naturally guide the visitors into the living room. As for the private spaces, natural lights are also guided into the rooms through the use of sliding doors and glasses. These streams of lights in a way give off a sense of freedom. Even though the spaces are partitioned into different zones caused by the original structural layout, but the abundant supply of lights give an illusion of a larger and freer space.

Furthermore, the clever use of materials also creates a warm and comfortable environment. The designer uses aged juniper wood to make the storage cabinet; he also uses the processed old sofa and aged tea wood, which are friendly towards both the visual and olfactory senses, to create a soothing environment. In addition, through the careful selection and integration of contour lines, the ceiling and the floor are made to complement well with each other, which in term creates a “smooth rhythm” of the design. Finally, the materials used also match well with the light source, which helps to build up an extremely stable and calm environment, allowing “broadness” to enter into the hearts of the house owners!

House Data

空间案名	台北卓宅
设 计 者	许华山
参与设计	许志彦
设计公司	许华山建筑师事务所
空间地点	台北市
空间性质	电梯大楼
空间面积	楼下层180m²、楼上层90m²、户外平台60m²
空间格局	玄关、客厅、厨房、餐厅、主卧室、主卧浴室、主卧更衣室、小孩房×2、客浴×2、储藏室、佣人房、起居室、琴房
主要建材	石材、抛光石英砖、玻璃、金属构件、再生柚木、再生桧木、牛皮再制、屋顶隔热板、废弃(再利用)树枝

Case Name / Taipei Juo's Apartment
Designer / Hua-Shan Hsu
Co-designer / Chih-Yen Hsu
Company / Hua-Shan Hsu Architectural Design Firm
Location / Taipei City
Nature of the apartment / Condominium
Area / Lower Level 180m², Upper Lever 90m², Outer platform 60m²
Outline / entrance, lounge, kitchen, dining room, master bedroom, master bathroom, master dressing room, kid's roomx2, guest's bathroomx2, storage, servant's room, living room, piano room
Main building material / stone, polished porcelain tile, glass, metal components, regenerated teak-wood, regenerated cypress, regenerated leather, roof insulation board, regenerated tree twigs.

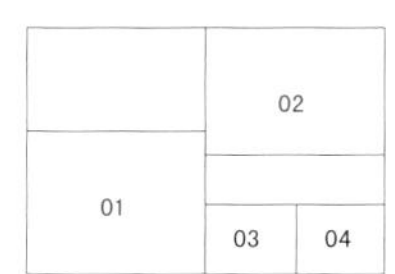

01 _ 客厅　02 _ 客厅　03 _ 玄关　04 _ 书房
01 _ living room　02 _ living room
03 _ entrance　04 _ study room

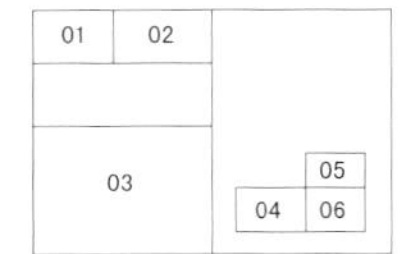

01 _ 餐厅　02 _ 主卧室　03 _ 走道　04 _ 餐厅　05 _ 小孩房　06 _ 小孩房
01 _ dining room　02 _ master bedroom　03 _ aisle　04 _ dining room
05 _ kid's room　05 _ kid's room

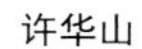

许华山
现　任／
许华山建筑师事务所 主持建筑师
学　历／
台北科技大学 建筑硕士
经　历／
台北市建筑师公会第十二、十三届学术委员会委员
台北市建筑师公会第十五届教育委员会委员
台北市小区建筑师
台湾室内装修学会 监事
台北科技大学设计学院创意学士班、金门大学建筑系 兼职讲师

Hua-Shan Hsu
Current Position
Chief Architect of Hua-Shan Hsu Architectural Design Firm
Education
Master of Architecture, Taipei University of Technology
Experiences
12th, 13th Academic Committee Member of Taipei Architects Association
15th Education Committee Member of Taipei Architects Association
Taipei Community Architect
Supervisor, Chartered Association of Interior Designers Taiwan
Adjunct Instructor of Undergraduate Program of Creative Design, National Taipei University of Technology, and Department of Architecture, Quemoy University

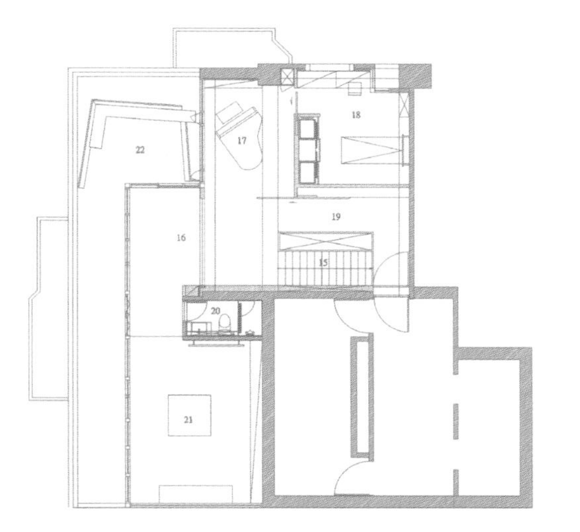

现代简约的两代大宅

A large flat in two modern minimalist styles

living room 客厅

屋主夫妻与成年未婚的儿子对于空间风格有着各自的喜好，设计师在经过与屋主多次沟通后，决定将两代人喜爱的风格融合，打造气派与品位兼具的大宅。

面积300m²的大宅，扣除公摊面积后室内大约230m²，如何展现出大宅的气势也是屋主所期待的。在设计中，设计师特意放大玄关空间的尺度，并巧妙地将琴室置入，此外还规划了展示柜，以及屋主期待的壁炉。这样的设计不但不违背主体风格，更展现出居住者的品位。客厅连接餐厅，以半开放式电视柜作为介质，连接厨房的夹丝玻璃对开拉门及书房的半开放玻璃隔断，延伸了空间。超乎面积的空间感，让人一进门就感受到空间的不凡气势。

为了展现空间质感，空间以大理石为主材质，不管是从电视墙延伸至书房的雪白银狐大理石墙面，或是镶着雪白银狐大理石边的卡布基诺大理石地板，还是以热带雨林大理石装点的沙发背景墙，大理石自然的纹理及温润的质感，搭配独一无二的天花板造型设计，让空间更显高贵优雅。

The residents of this unit are a married couple running an engineering business and their single son, and all of them have specific interests and preferred style for their personal space. After several discussions, the project designers decided to combine the different tastes into one, creating a flat that shows both quality and class.

However, displaying a myriad of styles while retaining the atmosphere of spacious quarters in a flat that's nearly 100m², with less than 70m² after taking away the facility rooms, is also part of the owners wishes and a challenge.

Hence, we deliberately enhanced the dimensions of the front porch, putting the piano room within. The display closet is to be combined with the fire place the owners wish to install, exhibiting the owners' unique taste without affecting the overall style. The living room is combined with the dining room with a semi-opened television cabinet acting as the semi-border. The opposing silver-lined glass sliding doors leading to the kitchen and the semi-opened glass enclosure of the study extends openness beyond the limits of the floor space. Once inside, the visitors cannot be feel the spacious nature of this house.

To further enhance the spacious quality of the flat, marble is adopted as the main material. The natural pattern and warm texture of marble combined with the most delicate finishing and novel ceiling design gives an elegant and classy appearance to the space, be it snow fox white marble extending from the television wall to the study, the cappuccino marble mixed with snow fox white lying beside, or the rainforest marble lining the back-wall behind the couch.

living room 客厅

living room hallway 客厅望向走道

Photo provided_IS International Design
图片提供_IS国际设计

House Data

设 计 者 陈嘉鸿
设计公司 IS国际设计
空间地点 新北市
空间性质 大楼
空间面积 300 ㎡
空间格局 玄关、客厅、餐厅、厨房、和室、次卧、次卧浴室、客用浴室、主卧房、主卧更衣室、主卧房浴室、观景阳台、工作阳台
主要建材 卡布基诺大理石、雨花白大理石、山水金峰大理石、热带雨林大理石、黑金峰大理石、雪白银狐大理石、南非黑大理石、咖啡绒大理石、铁刀木木皮、夹丝玻璃、茶色镜、黑玻璃、进口壁布、进口皮革、意大利进口瓷砖、进口海岛型木地板

Designer / Chia-Hung Chen
Company / IS International Design
Location / Taipei
Type of space / Large flat
Area / 300 m^2
Layout / Front porch, living room, dining room, kitchen, washitsu, second bedroom, second bathroom, guest bathroom, master bedroom, master changing room, master bathroom, viewing balcony, work balcony
Main materials / Cappucino marble, splattered white marble, mountain stream gold-peaks marble, rainforest marble, black gold-peaks marble, snow fox marble, south african black marble, coffee foam marble, senna bark, silver-lined glass, tea-colored mirror, black-tinted glass, imported wallpaper, imported leather, italian tiles, imported island wood flooring

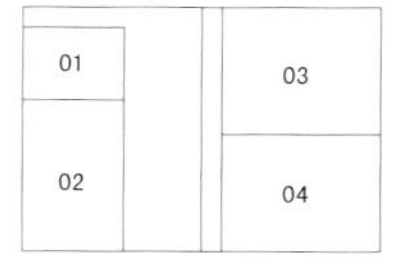

01 _ 餐厅 02 _ 餐厅 03 _ 餐厅望向客厅 04 _ 和室
01 _ dining room 02 _ dining room
03 _ dining room to living room hallway
04 _ washitsu

Designer Data

陈嘉鸿
现　任／
IS国际设计　主持设计师
简　介／
凭借天生的空间感和独特的美学品味，投入室内设计的领域，擅于运用垂直及水平线条来捕捉空间向度，并常以隐藏式的收纳方法及家具的搭配，打造出独特的居家风格。
著作／
《室内设计师陈嘉鸿的隐藏学》

Designer Data
Chia-Hung Chen
Current employment
Lead Designer, IS International Design
Resume
Chen combines an innate acute spatial sense with a unique set of aesthetics, and is fully devoted to the field of interior design. His precise handling of scale and skilled use of vertical and horizontal lines allows an accurate depiction of spatial direction. The hidden style of storage methods combined with th e furniture creates a unique residential style.
Publications
Hiding Methods of Interior Designer Chia-Hung Chen, published by MyHomeLife

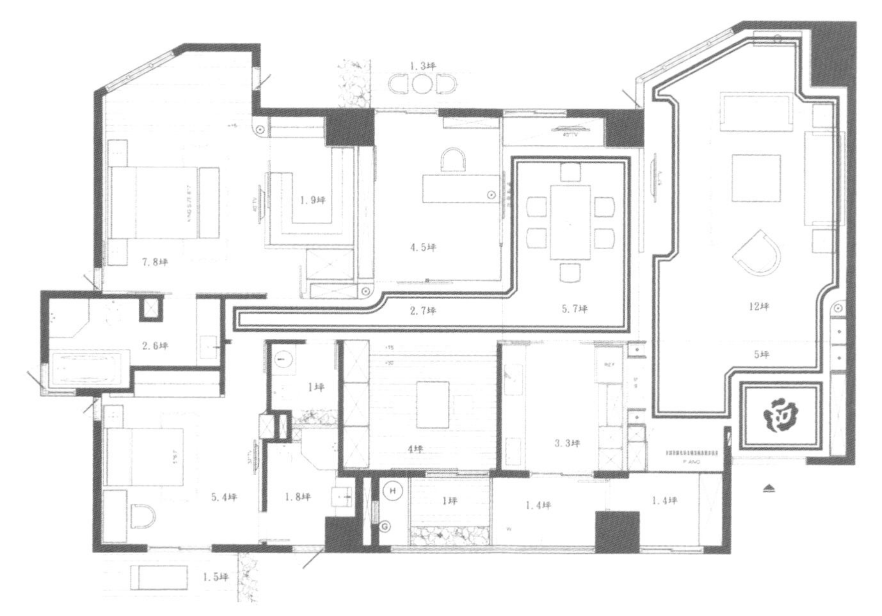

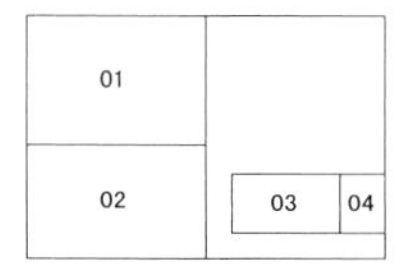

01＿和室　02＿主卧房　03＿卧房　04＿浴室
01＿washitsu　02＿master bedroom
03＿bedroom　04＿bathroom

优雅轻盈，美式古典大宅新诠释

Elegant and carefree, the new interpretation of an American classical mansion

living room and dining room 客厅与餐厅

屋主对于古典风格中简洁而优雅的设计非常喜爱，买下这个大宅，期望的就是给全家人一个舒适的居住空间。因此，设计师通过空间的重新定位，让格局符合屋主需求，并通过风格设定，让大宅成为屋主向往的居所。

通过精细的布局手法，私人空间与公共空间都有了合适的配置。比如玄关的设计，就以线板的层次感创造丰富的视觉效果，画框的配置也界定垂直的向度，搭配水晶灯与间接灯光，更表现出大宅的气势。而客餐厅采用开放式设计，并以统一的设计语汇，如线板、象牙色系等，营造出美式古典的优雅氛围，但又少了过于繁复的装饰。

餐厅的设计善用镜面进行处理，餐柜中间通过镜面材质的线条，将闪耀而具流动的视觉效果精准表达，再加上水晶灯、菱形切割镜面的使用，营造出气度不凡的用餐空间。通过设计语汇的掌握，以及结合功能性的设计，让大宅空间的优雅气质自然流露，以轻盈的姿态展现豪宅的另一种样貌。

The house owner has a strong liking for the minimalistic and elegant designs of the classical style. When he bought this property, his wish was that a comfortable living space is provided for his whole family. For this reason, the spatial layout of the apartment has been re-structured, so that the layout better suits the owner' s requirements. Furthermore, the apartment is also designed to suit the owner' s styling preferences.

Through the employment of intricate design techniques, both the private area and the public area are assigned to their proper configuration. For example, in terms of the design for the entrance, the layered effect of the line board has created a rich visual effect. As for the living room and the dining room an open design is employed. Furthermore, a unified design approach is adopted, for example, the use of line boards and ivory colored decorations all help to create a classic American-style elegance, but at the same time takes away the over-complicated decorative effect.

Mirrors play an important part in the design of the dining room. Through the line processing of the mirror material at the middle of the dining cabinet, a sparkling and fluid visual effect is created. Coupled with the use of chandelier and diamond-cut mirrors, a sophisticated dining space is created. By combining functionality and the stylish design, the designer has created an elegant and luxurious home, where a light-hearted approach has become another possibility.

Pictures_Chiang's Interior Design Ltd.
图片提供_蒋氏设计

dining room 餐厅

House Data

空间案名　中悦彼德堡
设 计 者　蒋学宏、蒋学正
参与设计　叶高志、温凯婷、张文彦、彭雅婷、吴佳怡、宋明璇、刘又绫
设计公司　蒋氏设计
空间地点　台北县三峡
空间性质　电梯大楼
空间面积　298 m^2
空间格局　玄关、客厅、餐厅、长辈房、小孩房×2、客浴、主卧室、主卧浴室
主要建材　石材、烤漆、马鞍皮、染布、雕花褐色镜、涂装木皮板、染白地板、铁件

Case Name / Jungyue Petersburgh
Designer / Hsueh-Hung Chiang, Hsueh-Cheng Chiang
Co-designer / Kao-Chih Yeh, Kai-Ting Wen, Wen-Yen Chang, Ya-Ting Peng, Chia-Yi Wu, Ming-Hsuan Sung, You-Ling Liu
Company / Chiang's Interior Design Ltd.
Location / Sanshia Town, Taipei County
Nature of the apartment/Condominium
Area/298 m^2
Outline / entrance, living room, dining room, elder's room, kid's room x2, guest's bathroom, master bedroom, master bathroom
Main building material / stone, enamel paint, Saddle leather, dyeing cloth, brown mirror with flower carvings, KD floor, white-dyed floor, cast iron.

01 _ 走道　02 _ 小孩房　03 _ 餐厅　04 _ 餐厅
01 _ aisle　02 _ kid's room　03 _ dining room　04 _ dining room

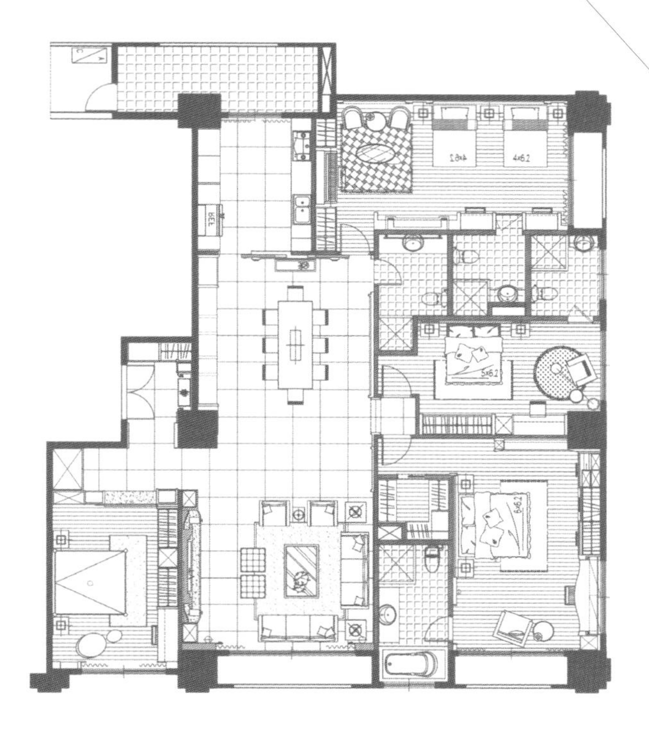

Designer Data

蒋学宏
现　任／
蒋氏设计 主持人
学　历／
中原大学建筑系毕业

Hsueh-Hung Chiang
Current Position
Chiang's Interior Design Ltd.Supervisor
Education
Master of Architecture, Chung Yuan Christian University

01＿长亲房　02＿主卧室　03＿玄关　04＿长亲房
01＿elder parents' room　02＿master bedroom　03＿entrance　04＿elder parents' room

蕴含东方人文价值的私宅

A house with eastern humanistic values

living room and dining room 客厅与餐厅

在看似西式的设计中，其实也能蕴含着无限的东方价值。在本案里，身为教授的屋主期望居家空间能融合美式古典的大方以及中国传统重视家庭的核心价值。通过空间格局的精准定位，每一个空间都有其承载的意义与功能。家在此处不是华丽的炫耀品，而是能包容居家成员的温暖场所。

考虑屋主的年龄、背景以及人文素养，本案的每一个空间都赋予特殊的意义。公共空间首重采光与通透性，干净的质地来自于玻璃屋光线的引入，以及楼梯间、电视墙后方两道光线的会合。视线受光线的引导，而在空间中来去自如，因而达到让空间放大的目的。

本案对奢华有着不同的定义：奢华来自于能和家庭成员共享的特别空间。二楼的书房为所有的居住成员量身打造，可以看到东方语言的诠释。阅读桌让全家人能在此处共享书香。通过书柜区隔的另一端空间，则可让父母亲舒适地处理事务。空间无论在功能性或是人性化的设定上，都能满足需求。至于私人空间的设计，则讲求私密性的营造，挑空的天花板放大了卧房的尺度，让空间得以延伸，并因此更显宽阔。

The seemingly Western design can be in fact, filled with boundless Eastern values. In this design case, the house owner, being a professor, wishes to combine the generousness of the classical American style with the Chinese's traditional emphasis on family values. Through a careful planning on the spatial layout, every space is endowed with its exclusive functionality and meaning. For this reason, home is no longer an extravagant place to show off, but a warm place to accommodate every members of the household.

Considering the house owner's age and his intellectual background, every space is given its special meaning. Special emphasis is placed on the lighting; natural lights from the conservatory and the two streams of light joined together behind the TV wall at the staircase allow for a cleaner outlook. Through the use of light, the purpose of visually magnifying the space is fulfilled.

In this case, the word "luxury" has been redefined. Luxury comes from the fact that each family member can come together at a specially built shared space. The study room at the second floor is tailored made for all the family members. From this, it is evident that Eastern family value has come into play. The reading table provides a place where every members of the family can come here to read. The parents can also do their official work at another side of the room as separated by the book shelf. Based on the above designs, the demands for functionality as well as user-friendliness are both met. As for the design of the private space, privacy is most emphasized. The heightened ceiling at the bedroom also allows for the visual extension of the space, giving an illusion of a larger room.

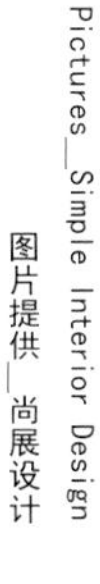

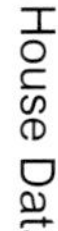

空间案名　青山镇郑宅
设 计 者　吴启民
设计公司　尚展设计
空间地点　新北市
空间性质　别墅
空间面积　297 m²
空间格局　车库、玄关、客厅、厨房、餐厅、佛堂、阅读区、书房、小孩房、起居室、主卧室、主卧更衣室、主卧浴室、客浴
主要建材　抛光石英砖、柚木地板、进口壁纸、文化石、实木天花板

Case Name / Chingshan Town Jeng's Property
Designer / Chi-Min Wu
Company / Simple Interior Design
Location / Taipei County
Nature of the apartment / villa
Area / 297 m²
Outline / garage, entrance, lounge, kitchen, dining room, ancestor worship altar, reading area, study room, kid's room, living room, master bedroom, master dressing room, master bathroom, guest's bathroom
Main building material / polished porcelain tile, teak-wood floors, imported wallpaper, great stone, wood ceiling

01	03
02	04

01 _ 餐厅与客厅　02 _ 阅读区与书房　03 _ 阅读区　04 _ 客厅
01 _ dining room and living room　02 _ reading area and study room
03 _ reading area　04 _ living room

Designer Data

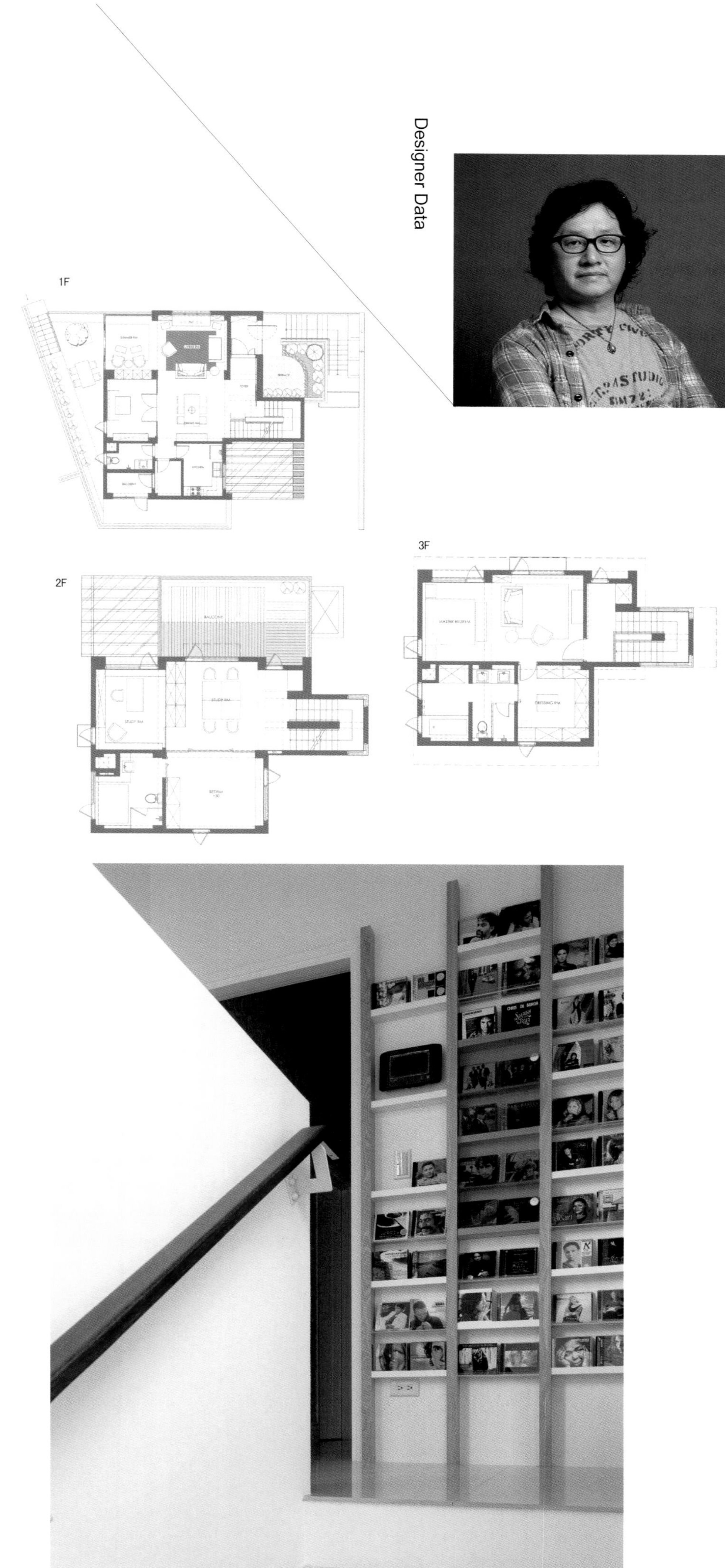

吴启民
现　任／
尚展设计 设计总监
学　历／
加拿大英属哥伦比亚大学（UBC）建筑与都市计划学系毕业
经　历／
汇侨设计、展亦设计 设计总监
2008年荣获AID Award 台湾室内设计大奖
2009年AID Award 台湾室内设计大奖入围
简　介／
始终坚持要利用房子本身的特性规划，顺着房子的优点来设计。有形的功能，回归自然变得纯粹。无形的价值，就不会被时间的潮流所淹没。奢华最极致的表现是融合自然与科技。着墨于每个微小的细节，让居住者的身心都能获得无比的满足与自在，这正是当前品味之士所追求的。

Chi-Min Wu
Current Position
Chief Design Officer of Simple Interior Design
Education
Bachelor of Architecture and Landscape Architecture, The University of British Columbia
Experiences
Chief Design Officer of Rich Honour Design Group and United Design Associate,
2008 AID Award
2009 Nominated AID Award
Brief
When designing for an apartment, I always strive to take the advantage of the original characteristic of the house. I insist to keep the good points intact. Functionality with forms allows the nature to become purer, so that values without forms would not disappear amongst the tide of time. The ultimate luxury is to combine technology and nature. I mind every single detail, so that both the mind and body of my clients can be satisfied and contended, the two values in which people with good taste would pursue.

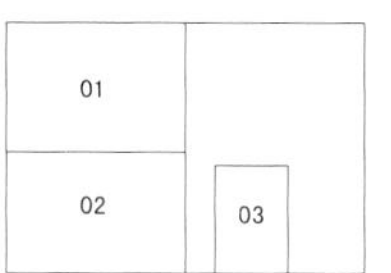

01＿主卧室与起居室　02＿主卧室　03＿主卧室入口
01＿master bedroom and living room
02＿master bedroom　03＿master bedroom entrance

living room 客厅

打造数字化的奢华大宅

Create a digitalized apartment

在都会大宅中如何能让古典的风格有完美的呈现？设计师在本案中，保留了最原始的古典元素，将简化了的美式古典作为设计主轴。经过全面性规划后，引进住宅计算机数字化管理系统，让豪宅在具有智能化功能的同时，也有气派奢华的古典样貌，展现住宅最精致的一面。

玄关入口为整个住宅的中轴线，通过圆拱的设计，将古典建筑的语汇转化至廊道，点缀出古典元素的意象。通过展示柜的区隔，不但让空间格局更具层次感，也让屋主的收藏品有合适的展示场所。客厅的主墙镜面饰以施华洛世奇的水晶，在低调中展现出奢华感。室内的家具均由加拿大进口，以求所有的软装饰与整体风格相呼应。

在本案的设计中，将智能化的设备融入空间。作为现代化的大宅，空间不但处处体现奢华与高质感，与最新科技的接轨也是设计的重点。用数字化的设备，让屋主即使在海外也可随时了解家中状况。

How do you retain the classical styles in an urban apartment? In this case, the designer Tung-Chou Chao decides to keep the classical elements intact, and uses the simplified version of the "American classical style" as the main emphasis of his design. After a comprehensive planning, digital computer management is introduced, so that the apartment can be modern and functional but also sophisticated in its classical exterior.

The entrance acts as the mid-point of the whole apartment. The use of arched design helps to bring a classical atmosphere to the corridor. Furthermore, by using the display cabinet as a divider, not only does it create a sense of layering, but also provides a place to exhibit the owner's treasured collections. The main wall mirror of the living room is decorated with Swarovski crystal, which is luxurious in a toned-down way. All the furniture used in this apartment was imported from Canada, so that the intended style can be rightly presented.

In this property, high-tech devices are hidden within the design. Being a modern apartment, not only has it got a touch of elegance and sophistication, but also stays in touch with the most recent technologies. Through the incorporation of digitalized facilities into the design, the house owner is able to monitor his house even when he is abroad.

Pictures_Dehan International Interior Planning Ltd.
图片提供_大涵国际室内企划有限公司

living room 客厅

House Data

空间案名　天母国堡陈公馆
设 计 者　赵东洲
参与设计　赵晏良
设计公司　大涵国际室内企划有限公司
空间地点　台北天母
空间性质　电梯大楼
空间面积　283 m²
空间格局　三房三厅三卫
主要建材　晶石、大理石、木作、乳胶喷漆、施华洛世奇水晶灯

Case Name / Tienmu country residence Chen's property
Designer / Tung-Chou Chao
Co-Designer / Yen-Liang Chao
Company / Dehan International Interior Planning Ltd.
Location / Tienmu, Taipei City
Nature of the apartment / Condominium
Area / $283m^2$
Outline / bedrooms x3,living rooms x3, bathrooms x3
Main building material / spar, marble, wood, latex paint, Swarovski Crystal

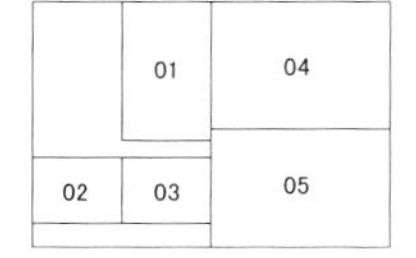

01 _ 廊道　02 _ 餐厅　03 _ 餐厅
04 _ 餐厅　05 _ 外玄关
01 _ corridor　02 _ dining room　03 _ dining room
04 _ dining room　05 _ outer entrance

Designer Data

赵东洲
现　任/
大涵国际室内企划有限公司设计总监
鸿室空间设计有限公司设计总监
台湾室内装修专业技术人员学会副理事长
台湾室内设计装修商业同业公会全国联合会理事
台北市室内设计装修商业同业公会常务理事
中原大学、中原大学研究所专家级助理教授
台湾建筑物室内设计技能检定命题委员
媒体专栏作家
学　历/
中原大学室内设计系硕士
经　历/
大仁室内计划公司项目经理
台湾装修技术人员协会秘书长
台湾室内设计协会理事
台湾中国科技大学室内设计系兼职讲师
华夏技术学院室内设计系兼职讲师

Tung-Chou Chao
Current Position
●Designer Executive of Dehan International Interior Planning Ltd. Company
●Designer Executive of Hong-Shih Interior Design Ltd. Company
●Vice Board Chairman of Chinese Association of Interior Design
●Board Chairman of Taiwan Association of Interior Design
●Managing Director of Taipei Association of Interior Designers
●Assistant Professor of Chung Yuan Christian University and Research Institute of Chung Yuan Christian University
●Examiner, Interior Design Skill Assessment, Council of Labor Affairs of Executive Yuan
●Columnist
Education
●Master of Interior Design, Chung Yuan Christian University
Experiences
●Project Manager of DAZA Interior Planning Co., Ltd.
●Secretary of Chinese Association of Interior Designers
●Director of Chinese Society of Interior Design
●Adjunct Instructor of Interior Design, Chinese University of Technology
●Adjunct Instructor of Interior Design Hwa Hsia Institute of Technology

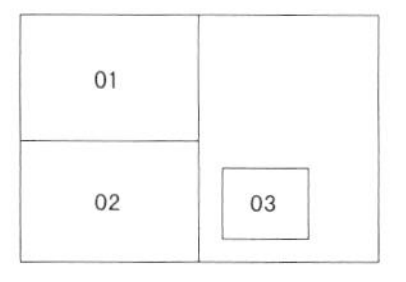

01＿主卧室　02＿小孩房　03＿书房
01＿ master bedroom　02＿ kid's room
03＿study room

living room 客厅

entrance 玄关

无疆界，给大宅一个新的定义

Without borders, the new definition of a mansion

别墅空间可以因其楼层设计而有更多的变化。也许常规的空间已让人厌烦，那么看看这个案子！在本案中，设计师试着将不可能之事变为可能——将树种在家里！

在此案中，建筑物为六层楼的别墅。除了通常的空间规划，最迷人之处在于屋主也可在自己的家中拥有一棵树。设计师通过垂直线的确立，在四楼的庭院种树，由此打破空间的固有框架，创造出新的可能。设计师表示，将界限拿掉，把生态"放"进来，是种树最主要的目的。在四楼以上的空间，将因为树的存在而有不同的视野，当然，居住的氛围也就跟着改变。在此处，大宅所包含的意义不只在其"大"，而在其"广"，因为通过视野的延伸，界限被打破了，广度自然也就展开。

在本案中，居住者的生活空间通过楼层来定义。客、餐厅位于同一层楼，让空间的公共性质更为明显。书房与吧台则介于卧房与客、餐厅之间，可视访客人数来决定是否开放，也为家人多创造一个休憩场所。通过广度的拓展，以及对空间的严谨定义，让空间的层次更为丰富。

There can be many variations in designs, depending on the floor level you are on. It is true that many people have already got tired of a conventional design. If you are one of them, then look no further! In this case, the designer has done the mission impossible: plant a tree in a house!

Apart from the somewhat rigid spatial layout, what is most charming about this six-storied villa is that the house owner is able to own a tree inside his house. Through the establishment of four vertical lines, a tree is planted on courtyard at the 4th floor. Through the planting of this tree, the rigid framework of the space is broken, creating new possibilities. The main purpose of planting the tree, as stated by the designer, is to remove boundaries and bring ecology in. Due to the presence of the tree, the spaces above the 4th floor will exhibit a different spectacle. Of course, the whole living atmosphere will change, too. In here, the meaning of a villa does not lie on its "largeness", but on its "broadness" instead. Through the employment of visual extension, boundaries are broken, and the broadness comes naturally.

In this case, the living space of the inhabitant is defined by the floor levels. The living room and the dining room are both located on the same floor, making the openness of the public space more evident. As for the study room and the bar, they are situated in between the bedroom and the living and dining room. These spaces can be both opened or closed, depending on the number of visitors. Furthermore, these places also act as a leisurely space for the family members. Through visually expanding the horizontal axis, as well as redefining the space, the quality of living for the inhabitants has definitely moved up to the next level.

Pictures_Chiang's Interior Design Ltd.
图片提供_蒋氏设计

dining room 餐厅

House Data

空间案名 墅无界
设 计 者 蒋学宏、蒋学正
参与设计 叶高志、温凯婷、张文彦、彭雅婷、吴佳怡、宋明璇、刘又绫
设计公司 蒋氏设计
空间地点 新竹县竹北市
空间性质 别墅
空间面积 265 m²
空间格局 客厅、餐厅、书房、吧台、主卧室、主卧浴室、水疗区、客房、客浴
主要建材 洞石、竹片、木皮、黑色玻璃、卡拉拉白大理石、橡木拼板、香杉木、水泥

Case Name / Villa Unbounded
Designer / Hsueh-Hung Chiang, Hsueh-Cheng Chiang
Co-designer / Kao-ChihYeh, Kai-Ting Wen, Wen-Yen Chang, Ya-Ting Peng, Chia-Yi Wu, Ming-Hsuan Sung, You-Ling Liu
Company / Chiang's Interior Design Ltd.
Location / Jhubei City,Hsinchu County
Nature of the apartment / villa
Area / 265m²
Outline / living room, dining room, study room, bar, master bedroom, master bathroom, SPA area, guest's room, guest's bathroom
Main building material / Travertine, bamboo slices, black glass, Bianco Carrara marble, oak laminated floor, Cedar, concrete paint.

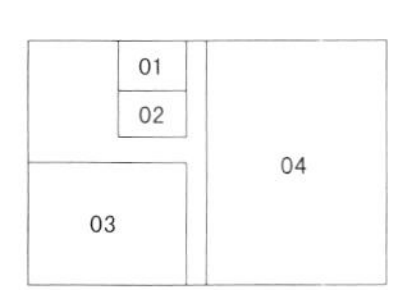

01 _ 吧台 02 _ 楼梯间 03 _ 书房 04 _ 庭院
01 _ bar 02 _ staircase 03 _ study room
04 _ courtyard

01	02	03	
04			05

01 _ 水疗区　02 _ 浴室　03 _ 主卧室　04 _ 水疗区　05 _ 主卧室
01 _ SPA area　02 _ bathroom　03 _ master bedroom
04 _ SPA area　05 _ master bedroom

Designer Data

2F

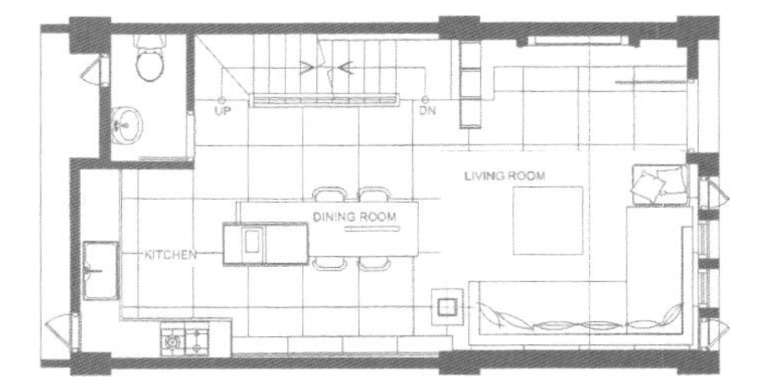

4F

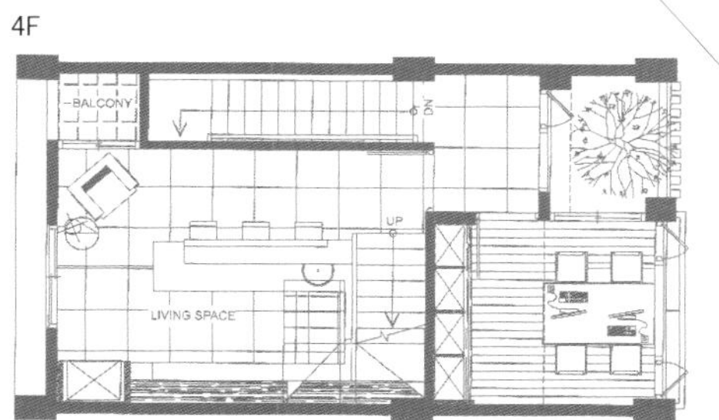
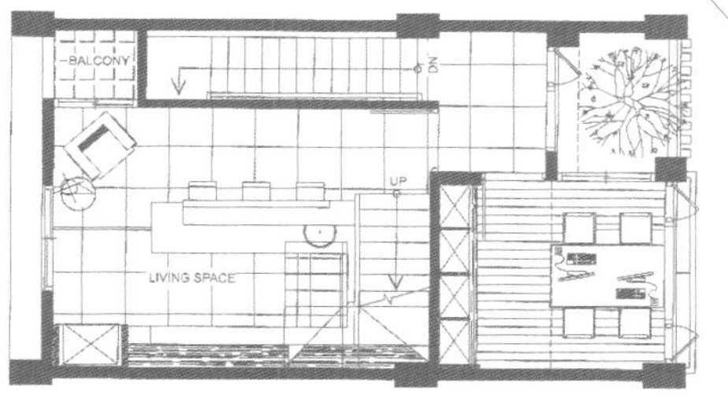

5F

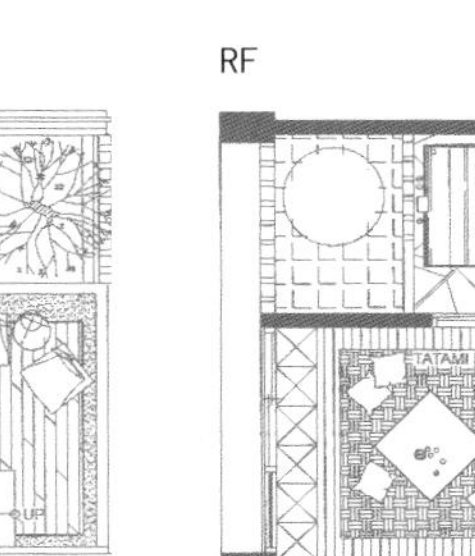

RF

蒋学宏

现　任／

蒋氏设计 主持人

学　历／

中原大学建筑系毕业

Hsueh-Hung Chiang

Current Position

Chiang's Interior Design Ltd.Supervisor

Education

Master of Architecture, Chung Yuan Christian University

低限时尚，利落大宅

Low-key fashion, neat and spacious apartment

living room 客厅

对质感的追求是本案设计的重点。为了让空间的大气与宽阔感能有最完美的呈现，设计师在公共空间的设计中，不但将时尚的设计手法带进细节之中，同时也善用建材与家具的搭配，将“如画的”概念融入其中。一种时尚却不浮夸、利落却不简陋的设计，将都会居家个性做一全新的演绎。

玄关入口的隔屏，以黑色玻璃做大面积的切割构图，平面的雕塑感产生了层次分明的造型效果，并且呼应天花板造型的雕塑语汇。奢华的气质主要来自于地面仿皮革砖，其独特的纹理带出视觉变化，同时也兼具易清理的特点。电视墙虽以白色作为对比色，但是设计创造出纹理的趣味，并带来光影变化。至于沙发后方的墙面，通过大面积铺陈仿鳄鱼皮革砖，将时尚的主题表现无遗。

走道通过门片表面的切割设计，以不规则的几何形再次呼应设计主题。悬吊式的柜体结合灯光设计，不但减轻柜体重量感，也具有引导视线的作用。主卧室设计利用上下对称的弧线语汇，营造出空间的层次感，展现私人空间的舒适与个性。

The pursuit of quality is the main focus of this design. In order to perfectly show the grandness and broadness of the space, the designer not only incorporates many stylish techniques into his design, but also makes building materials and furniture match well with each other. The concept of being “picturesque” is embedded in his design. By looking at his design, the word “urban character” has suddenly found a new meaning: fashionable, but never flashy, minimalistic but never simple.

A room divider made of generously-cut black tinted glasses is placed at the entrance. It has a sculpted surface which creates a layered effect. This complements with the ceiling, with also has a sculptured surface. A sense of luxury comes mainly from the leathered-effect tiles on the floor. Their unique texture is both visually stimulating and at the same time easy to clean. Although the wall on which television is mounted uses white as the contrasting color, but on the front layer the designer uses some interesting textures, which bring out variations in light and shade. As for the wall behind the sofa, crocodile-leathered effect tiles are used. These tiles are used over a large area, which makes the whole atmosphere fashionable.

As for the aisle design, the surface of the door is cut into geometric shape to match the theme. The suspending cabinet incorporates a lighting design, which not only reduces the visual weight, but also has the function of visual guidance. Vertically symmetric design is used for the master’s bedroom, which create a sense of layers in the space, which in turn exhibits the coziness and uniqueness of the private space.

Photo provided_Yun-Yih Interior Design
图片提供_云邑设计

aisle 走道

living room 客厅

dining room and living room 餐厅与客厅

House Data

空间案名　三峡朴园
设 计 者　李中霖
设计公司　云邑室内设计
空间地点　新北市
空间性质　跃层
空间面积　260 m^2
空间格局　玄关、客厅、厨房、餐厅、起居室、主卧室、主卧浴室、主卧更衣室、男孩房、男孩房浴室、女孩房、女孩房浴室、女孩房更衣室、客浴、书房
主要建材　进口砖、木皮、磁漆、黑色烤漆玻璃、铁件

Case Name / Shan Shia Pu Yuan
Designer / Chung-Lin Li
Company / Yun-Yih Interior Design
Location / Taipei County
Nature of the apartment / Condominium
Area / 260m^2
Outline / entrance, living room, kitchen, dining room, living room, master bedroom, master bathroom, master dressing room, boy's room, boy's bathroom, girl's room, girl's bathroom, girl's dressing room, guest bathroom, study room
Main building material / import bricks, wood, enamel paint, black tinted glass, components made of iron

01 02 03 04

01 _ 厨房与餐厅　02 _ 餐厅与客厅　03 _ 厨房、餐厅与客厅　04 _ 餐厅与客厅
01 _ kitchen and dining room　02 _ dining room and living room
03 _ kitchen, dining room and living room　04 _ dining room and living room

Designer Data

李中霖
现　任／
云邑室内设计总监
学　历／
复兴美工室内设计组毕业
经　历／
1998 云邑室内设计工作室成立
2001 云邑室内设计有限公司成立

简　介／
不断地自我实践对设计的理想，总能随时与潮流接轨，以不拘泥的思维与精准落实的态度，在设计的旅程中向前迈进。

Chung-Lin Li
Current Position
Director, Yun-Yih Interior Design
Education
Fu-Hsin art school, interior design unit
Experiences
Founded Yun-Yih Interior Design Workshop in 1998
Founded Yun-Yih Interior Design, Ltd. In 2001

Brief
Always strive to realize my designing goals but in the same time catching up with the current trend. With an unconventional thinking and down to the earth attitude, I embrace my journey of interior designs ahead.

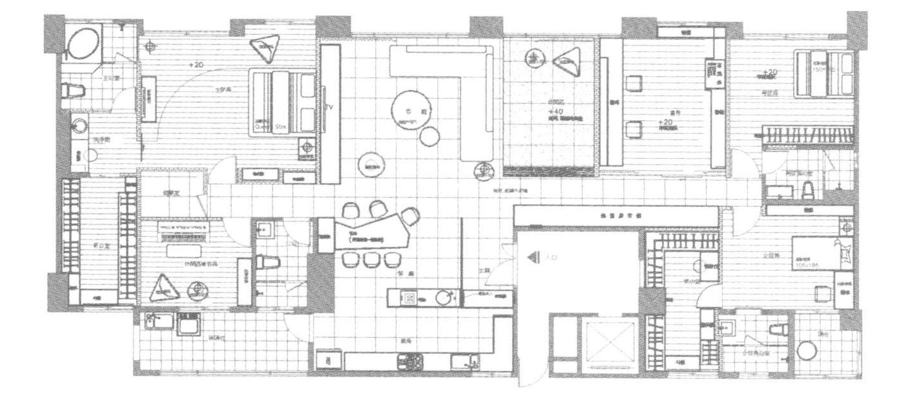

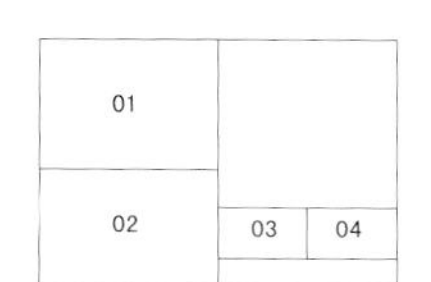

01 _ 厨房、餐厅与客厅　02 _ 玄关、餐厅与客厅
03 _ 书房　04 _ 主卧室
01 _ kitchen, dining room and living room
02 _ entrance, dining room and living room
03 _ study room　04 _ master bedroom

让家像城堡一样安全而舒适

Let your home be as safe and comfortable as a castle

living room 客厅

想象回到家，就如同进入私人城堡一般，能感受到安全、私密而温暖的氛围。本案为上下楼两户打通的设计案，为展现出大宅的气势以及稳重的格调，空间设计以新古典风格为主。通过公共区域与私人区域的合宜界定，为屋主打造出功能丰富的居住空间。

玄关入口刻意以圆形为主要框架，通过几何形本身的意涵带出家的圆满意象。局部的金色与线板元素带出空间的新古典氛围，呈现出细腻而又有些许奢华的质感，沙发背墙悬吊大型画作，强化了空间的视觉焦点。另一面墙则通过明镜、茶色镜与玻璃纸，创造出拼贴的趣味性，既可让空间感放大，也创造出视觉美感。

餐厅空间的设计将功能性融合其中，大面的拉门采用金属收边，银色墙面与水晶灯衬托出空间的华丽感，一旁的娱乐室利用玻璃隔断，让空间更为通透。二楼的私人空间强调私密性，楼梯的玻璃与墙面上的图案在设计语汇上相呼应，而以五星级饭店概念设计的过道，营造出安稳而舒适的风格，提升了屋主的生活质量。

Imagining going home is like going to a private castle, you can feel the safety, enjoy the privacy and a warm atmosphere. In this case, two floors are renovated into one duplex apartment. In order to fully show the quality of a large and sophisticated apartment, the designer decides to use classical style to decorate this property. By subtly divide the property into public and private zones, a functional living space is created.

A round-shaped frame has purposely been used at the entrance. The employment of geometric shape symbolizes a harmonious home. As you look closely into the detail, you can sense an element of neo-classical mood, as evident by the employment of gold-colored paints as well as smooth contour lines. This helps to create gentle textures with a hint of extravagance. On the wall at the back of the sofa a large painting is mounted. This helps to strengthen the visual focus. At another wall, transparent and opaque mirrors as well as cellophane are used as wall decoration; this creates something of a collage art. Not only does it creates a sense of an enlarged space, but is also visually entertaining itself.

Functionality has been incorporated into the design of the kitchen. Metal frames are used on the large sliding door; the use of silver paint on the wall complement with the chandelier above which bring out the beauty of the space. Glass panel is used as a divider for the recreational room besides the kitchen. This allows for a more transparent sense of space. Privacy is emphasized at the second floor where it is used as a personal space. Glassy materials on the stairs complement well with totems on the wall. Finally, the aisle that is designed to purposely resemble the style of a five-star hotel not only points out the owner’s yearning for a good quality of living, but also create a feeling of security and comfort.

Pictures_Interplay Space Design Studio
图片提供_演拓空间室内设计

House Data

空间案名	夏悠宫王宅
设 计 者	张德良、殷崇渊
参与设计	演拓空间室内设计
空间地点	台北市
空间性质	电梯大楼
空间面积	1楼 127 m² 、 2楼 127 m²
空间格局	客厅、餐厅、视听室、客浴、主卧室、主卧浴室、客浴、小孩房×2
主要建材	大理石、抛光石英砖、墨镜、烤漆玻璃、喷漆、壁纸、裱布、金属箔、画框、木地板

Case Name / Xia Yu Gong Wang's Home
Designers / Te-Liang Chang, Chung-Yuen Yin
Company / Interplay Space Design Studio
Location / Taipei City
Nature of the apartment / Condominium
Area / 1F 127m², 2F 127m²
Outline / living room, dining room, audio-visual room, guest's bathroom, master bedroom, master bathroom, guest's bathroom, kids rooms x2
Main building material / marble, polished silica brick, sunglasses, painted glass, spray-paint, wallpaper, mounted cloth, gold leaf, picture frames, wooden floor

01	03
02	04

01 _ 餐厅 02 _ 客厅 03 _ 客厅 04 _ 餐厅
01 _ dining room 02 _ living room 03 _ living room 04 _ dining room

Designer Data

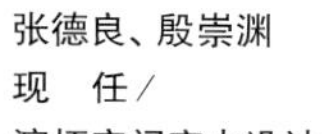

张德良、殷崇渊
现　任／
演拓空间室内设计 总监
经　历／
台北市室内设计装修商业同业公会会员
1999 创立演拓空间室内设计
简　介／
坚持好的设计应是体贴使用的人。设计的目的在于使生活简单，而不是让居住者花过多的时间整理照顾，不应该让不合宜的设计造成居住者的负担。

Te–Liang Chang, Chung–Yuen Yin
Current Position
Director, Interplay Space Design Studio
Experiences
Member of Taipei Association of Interior Designers
1999– Established Interplay Space Design Studio
Brief
A good designer should be considerate towards his/her client. The design aims to make life simpler, rather than letting the residents spending too much time taking care of their property; and should not cause burden for the inhabitants with inappropriate designs.

5F

6F

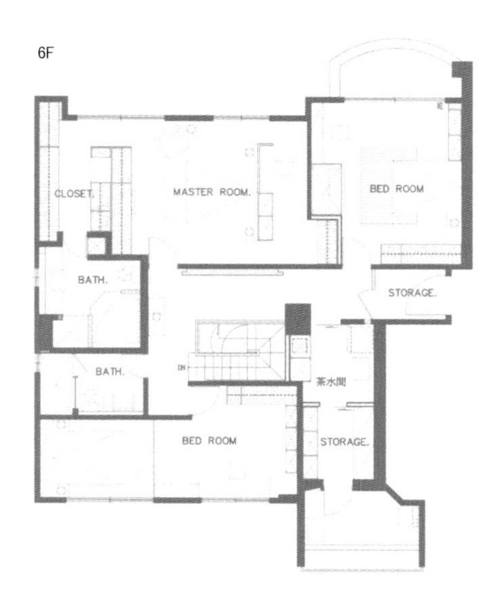

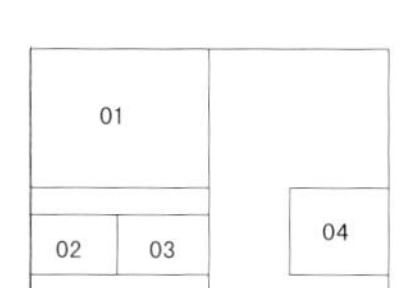

01 _ 主卧2　02 _ 2楼过道　03 _ 主卧1　04 _ 小孩房
01 _ master bedroom 2　02 _ 2F aisle　03 _ master bedroom 1
04 _ kids room

萃取，抽除立面
让空间还原
Restoring space by extracting the essence of objects

intermediate landing 梯间

dining room 餐厅

stairs 楼梯

如果说科技来自于人性，而所有的产品都朝着功能整合的方向进化，那么空间设计也不会只停留在单一向度。从这样的角度来思考，家居设计是否也有可能在常规的建筑框架下有新的发展？

站在入口望去，空间宽，视野阔，干净的线条、简单的结构，将居住空间还原至本质。大面积的采光，几乎无墙面的设计，可以看出设计师不过分装饰的初衷。为了将线条整合到最简洁，空间简化出数条水平线，地面、天花板，甚至连厨具的高度都经过精心设计。

除了线条收拢以及脉络化的整理，立面的解散也是此空间中最为精彩的设计。以一楼的客房为例，平常空间并不封闭，通过地板设计将空间延伸，空间仅以大面积的拉门区隔。二楼的私人空间同样如此，玻璃的使用让空间更为通透，墙面并不突出其装饰性，只有在需要区隔和保护隐私时，才以拉门的形式出现。至于一楼的客浴则更让人惊喜，看似清透的玻璃隔断，当有人使用时，液晶玻璃会让客浴空间变得不透光。科技手段的引入，功能被完整保留，而墙面却消失在空间中。

If technology is an invention of humanity, and if every product is evolving along the lines of combined functionality, then interior space design would not stay very long along this unidirectional path. Thinking along these lines, isn't it natural to conclude that residential interior designs can also create novel developments under the current architectural system.

Standing at the entrance, a wide and unbroken vista greets our view. Simple clean lines and minimalist structures have restored the meaning of residence back to its original essence. Large areas of natural lighting and a design with minimum compartmentalizing walls allow one to see the starting philosophy of the designers. In order to give the cleanest look to the spatial lines and create a few horizons along the space, the heights of the floor, ceilings and kitchen utensils have all been specifically redesigned.

Besides realigning the lines to give a network reconfiguration, the dissection of the compartments remains the most exciting design of this project. For example, the first floor guest room is left open at most times, separated and closed by using a large sliding door in the utilization of floor planning to extend spatial volumes. The private rooms in the second floor are also similarly designed, with the glass materials chosen for their visual transparency. The main function of wall surfaces is not only a provision of decor, but to provide privacy only when required, hence sliding doors are incorporated as a result. The first floor guest bathroom is another expression of this philosophy. When in use, the plasma glass will cause the pane to turn opaque, turning off visual access into the showering space. The integration of technology preserves function of the wall, yet causes it to disappear in real space, leading to increased spaciousness.

Pictures_Taibei Base Design Center
图片提供_台北基础设计中心

House Data

设 计 者　黄鹏霖
设计公司　台北基础设计中心
参与设计　ROY、CASSIE、KOBE
空间地点　台北市
空间性质　电梯大楼
空间面积　248 ㎡
空间格局　玄关、客厅、厨房、餐厅、主卧室、主卧更衣室、主卧浴室、小孩房、客用浴室
主要建材　复古面石材、银狐大理石、铁件、纯白压克力板、镀钛钢板

Designer / Peng-Lin Huang
Company / Taipei Basic Design Center
Participants / ROY, CASSIE, KOBE
Location / Taipei
Type of space / Large flat
Area / 248 m^2
Layout / Front porch, kitchen, dining room, master bedroom, master changing room, master bathroom, children's room, guest bathroom
Main materials / Revivalist stone facing, snow fox stone, rust resistant iron components, pure white acrylic, titanium-plated steel plates,

01	02	03
01	04	

01 _ 餐厅　02 _ 客厅　03 _ 客厅　04 _ 餐厅
01 _ dining room　02 _ living room
03 _ living room　04 _ dining room

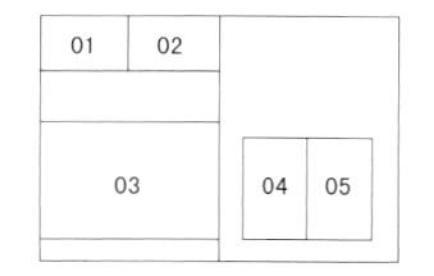

01 _ 楼梯　02 _ 书房　03 _ 浴室　04 _ 走道　05 _ 玄关
01 _ stairs　02 _ study room　03 _ bathroom　04 _ hallway　05 _ front porch

Designer Data

黄鹏霖

现　任/

台北基础设计中心主持设计师

简　介/

专注于常被人忽略的空间细节，期望在设计上能多一点想象力，多一点尝试，创造比预期还要多很多的生活乐趣。自成立设计中心以来获奖无数，曾获得第七届现代装饰国际传媒奖杰出设计师奖、TID（台湾室内设计大奖）空间家具奖、TID住宅空间奖等，并被评为美国室内设计杂志中文版十大封面人物以及最受欢迎设计师等。

Peng-Lin Huang

Current employment

Lead Designer, Taipei Basic Design Center

Resume

Huang focuses on spatial details that most tend to neglect, hoping that a bit more creativity and a bit more attempts can be made to create a livelier residential environment than the original plans. After starting his design center, Huang has won numerous rewards, winning outstanding designer in the 7th Modern Decoration International Media Prize, TID Spatial Furniture Award, TID Residential Space Award and featured as one of the top ten personalities in the American Interior Design Magazine-China and the most popular designer award.

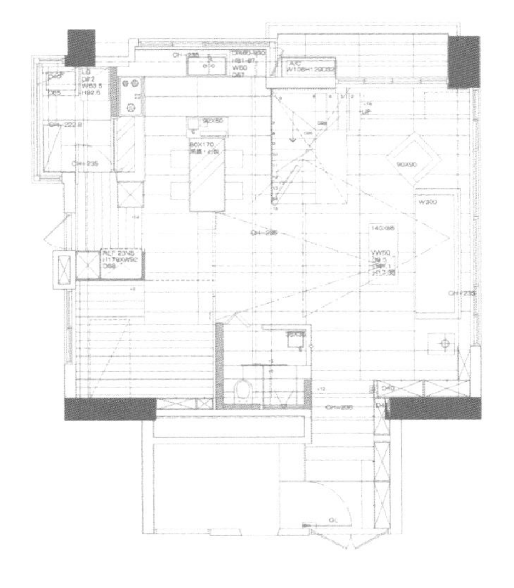

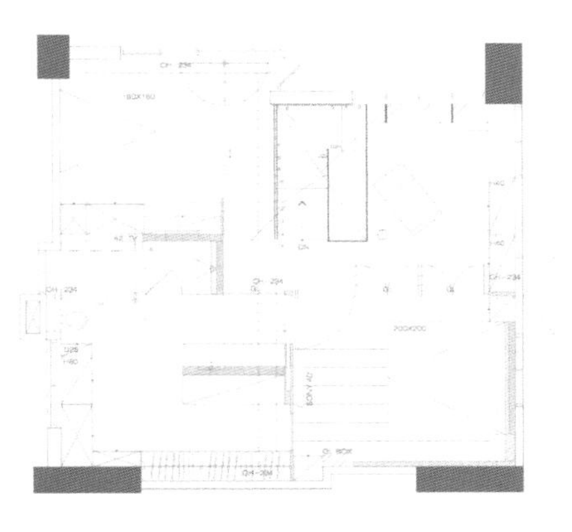

时尚奢华中展现高雅品味

Taste of luxury in the elegant fashion show

living room 客厅

living room 客厅

本案位于仁爱路上，是艺念集私为屋主夫妻设计的第三套房子。在前两次的合作中，屋主夫妻考虑到孩子年纪尚小，因此空间风格设定为温馨与简约，而这次的设计则希望能满足屋主夫妻的喜好。

从事时尚精品业的女主人，对于设计与美感有着极高的期望，希望居家可以呈现精致奢华的设计感。而男主人则对意大利现代简约风格情有独钟，因此家具单品大多由男主人亲自出马挑选。

简约与奢华，看似两种截然不同的格调，设计师是如何协调安排呢？设计师表示，男主人所挑选的意大利品牌家具多是简约风格，本身已经具有极高的美学品位，再利用许多配件与颜色，让色彩更加丰富，并带出奢华气度。例如施华洛世奇品牌水晶壁灯，通过光线折射而使空间更具耀眼光彩。以白为主的餐厨区中，设计师在中岛处设计了具有特别图案的玻璃灯箱，让空间变化更具创意。而在主卧室，则特别为女主人量身设计了整面的香水展示柜，不但实用而且相当华丽，让女主人可在此悠闲地享受一天的好心情。

Pictures_ D-A design Ltd.
图片提供_ 艺念集私

This property is located at Ren-Ai road, Taipei. It is the third apartment which D-A design Ltd has designed for the couple. In the previous two liaisons with the couple, minimalistic styles were employed, considering the fact that it was the couple's children which the properties were designed for, and minimalistic style suits their youth and vigor. However, in this case, it is the couple's apartment which is to be designed. For this reason, the design itself has to reflect the couple's own preferences.

The hostess works at fashion industry, for this reason, she has a very good taste in the area of design and aesthetics, in general. She wishes that the apartment design would reflect the refinedness and luxury which she holds high regards of. As for the host, he has developed a special liking for modern Italian minimalist style. For this reason, most of the furniture's are personally chosen by the host.

Simplicity and luxury: how exactly do the designers mix-match and coordinate these two seemingly different styles? The designers state that the minimalistic Italian furniture individually selected by the host is already high in artistic taste. The designers only need to make the overall coloration of the apartment more vivid in tonality by using more accessories and colors. By doing so, the luxurious of the apartment is fully exhibited. Examples include the use of Swarovski crystal chandelier, which displays its dazzling beauty through the light refractions it created; and the use of patterned light box made of glass at the mid-island area at the white-painted kitchen; also add a sense of creativity to the spatial design. As for design for the master's bedroom, a perfume display cabinet has been especially tailored made for the hostess; which is not only functional, but also very elegant, allowing the hostess to prepare a day in good mood within this relaxing fragrance.

House Data

空间案名 仁爱路 住宅案
设 计 者 黄千祝、张绍华
设计公司 艺念集私空间设计有限公司
空间地点 台湾台北仁爱路
空间性质 电梯大楼
空间面积 231 ㎡
空间格局 玄关、客厅、起居间、餐厅、开放式厨房、浴室、书房、主卧房、主卧浴室、二间小孩房
主要建材 香榭蓝大理石、激光切割铁件、茶色玻璃、镜子、窗帘、水晶灯、进口壁纸、金属漆壁板

Case Name / Ren-Ai Road apartment case
Designer / Anita Hwang & David Chung
Company / D-A design Ltd.
Location / Ren-Ai Road, Taipei, Taiwan
Nature of the apartment/Condominium
Area / 231m²

Outline / entrance, lounge, living room, dining room, open-designed kitchen, bathroom, study room, master bedroom, master bathroom, two kid's room
Main building material / Pallisandro Blue marble, laser-cut metal components, brown glass, mirror, curtain, chandelier, imported wallpaper, wall board with metal paint.

01 02 03 04 05

01 _ 客厅 02 _ 玄关 03 _ 客厅 04 _ 客厅 05 _ 主卧
01 _ living room 02 _ entrance 03 _ living room
04 _ living room 05 _ master bedroom

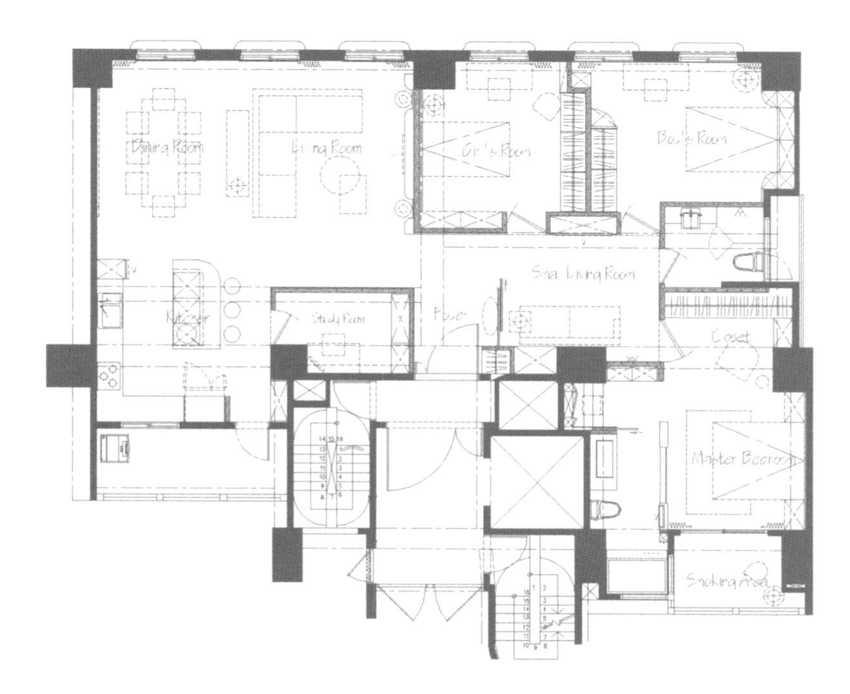

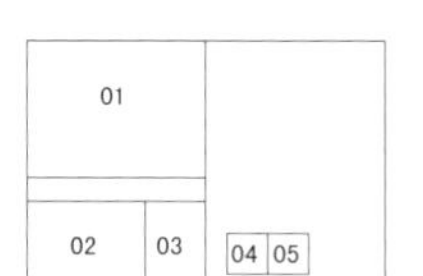

01 _ 男孩房　02 _ 主卧香水展示柜　03 _ 主卧更衣室
04 _ 主卧卫浴　05 _ 女孩房
01 _ boy's room　02 _ master bedroom perfume display cabinets
03 _ master dressing room　04 _ master bathroom　05 _ girl's room

Designer Data

黄千祝
现　任／
艺念集私空间设计有限公司　创意总监
学　历／
协和工商职业学校室内设计科毕业，美国普莱斯顿大学高级工商管理硕士

张绍华
现　任／
艺念集私空间设计有限公司　执行总监
学历／
复兴美工空间设计专业毕业

简　介／
从不设限自我风格，善用素材本身的美感，利用不同材质及配色展现空间变化。擅长使用建材的特性，特别专注于铸铁材质的运用，将其自然呈现于空间设计中。创意的混搭装置美学，营造出具有个人特殊风格及时尚的空间，让居住者得以释放压力，又呈现出空间的无限可能性。

Anita Hwang
Current Position
D-A design Ltd. Chief Creative Officer
Education
Unit of Interior Design, Xie-He High School of Industry and Commerce, Preston University / EMBA

David Chung
Current Position
D-A design Ltd. Chief Executive Officer
Education
Graduated in Division of Space Design, Unit of Art and Craft, Fu-Hsin Trade and Art School

Brief
Never limit oneself to a certain style, we are able to show off the natural beauty of the materials themselves. By using different materials as well as color complementation, the variation in the spatial design is fully exhibited. We are also good at taking advantage of the special properties of the materials themselves. A special emphasis is placed on the use of casted irons, and naturally presents them in our designs. Finally, a fashionable space is created where unique and personalized styles are valued by mixing creativity and aesthetics. This allows the inhabitants to release the tension in their lives as well as creating designs with infinite possibilities.

设计无受限，才是真实的设计

No style is my style!
True design only comes from the absence of limits.

hallway 走道

living room 客厅

受荷兰风格派核心人物杜斯堡（Theo van Doesburg）设计思维影响，设计师撷取其中艺术元素，将斜角经线面扩张，转化成有格调的旋律，再运用几何造型的规律及秩序性，让空间呈现等量分割的表情。此外，通过将绘画元素的风格转译，设计师将平面立体化、立体平面化，同一个设计语汇不断地流转在空间中的各个形式中，让整个居家空间为艺术灵光所包围，创造出丰富的层次。

雕塑语汇仿佛在墙上凿出一块块的几何体，又仿佛将音符立体化，跃动于墙面之上。平面化的立体形象，将风格派的画作重现，将"艺术"一块块地拼贴成型。双向走道底端，由皮革分割绷制而成的端景墙面，或书房收纳柜雕塑性的设计，让同一元素附着于不同的材质与立面上，从线、面发展到立体，笔直而流畅的线条，通过斜裁产生不对称的效果，串联出空间的犷达气势。

设计师将古典精神融入现代空间，让居住者感受到抽象的秩序与均衡。此外，主卧室与小孩房以浴室为中心做双动线设计，联动式的空间规划，成就了宽敞的浴室，也让设计改变生活的形式。

This project is most influenced by the concepts of Theo van Doesburg, the De Stijl designer whose neo-plasticism styles was most influenced by contemporary cubism. In this project, Wang takes the artistic elements of Theo and expanded the diagonal angles along linear surfaces, transforming the diagonals into a rhythmic pattern with a distinct geometric order and law. Using a language of painting to express spatial bisections, the designer inter-converts 2D planes and 3D space, adopting a single design language in a continuous flow across every form of space, surrounding the residential quarters within the inspirations of art and forming a rich dialogue between the layers of design.

The language of neoplasticism creates geometric figures whilst seemingly gives physical form to its melodies that the figures seems to be moving across the wall. Such is the effect of giving 3D characteristics to a planar surface, formulated by puzzling the artistic styles of De Stijl together. Other design features include the two way walkways which end with a leather covered end wall or a sculpted-style bookcase. By combining the same elements across different textures and surfaces, the straight lines can run diagonally across compartments and objects, extending across lines, surfaces and 3D objects to form a charming asymmetry, composing an unfettered theme of space.

By following old traditions and incorporating it into modern space, the residents are given a result giving them an abstract order and balance. Additionally, the master bedroom and children' s room are designed with two dynamic lines with the hub at the bathroom. The dynamic spatial planning created a spacious bathroom, becoming the perfect expression of how designs change lifestyles.

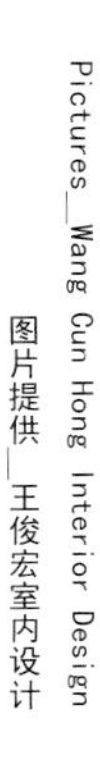

Pictures_Wang Gun Hong Interior Design
图片提供_王俊宏室内设计

House Data

设 计 者 王俊宏
设计公司 王俊宏室内装修设计
空间地点 台北市
空间性质 大楼
空间面积 230 m²
空间格局 客厅、餐厅、厨房、主卧室、长亲房、小孩房×2、卫浴×3
主要建材 木皮、铁件、茶色玻璃、石材

Designer / Cun-Hong Wang
Company / Wang Cun-Hong Interior Design
Location / Taipei
Type of space / Large flat
Area / 230 m²
Layout / Living room, dining room, kitchen, main bedroom, elder's room, children's room × 2, bathroom × 3
Main materials / Bark, iron components, tea-colored glass, stone materials, spray paint, tea-colored mirrors, teak flooring

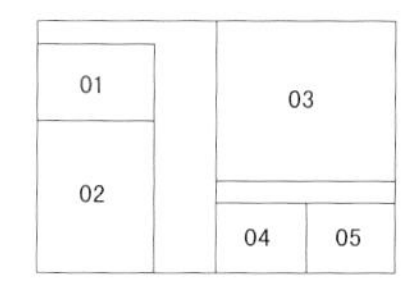

01 _ 客厅 02 _ 餐厅 03 _ 餐厅 04 _ 餐厅 05 _ 走道
01 _ living room 02 _ dining room
03 _ dining room 04 _ dining room 05 _ hallway

Designer Data

王俊宏
现　任／
王俊宏室内装修设计 主持设计师
简　介／
擅于表现居家空间的简约与时尚；调和对比色差使空间感放大；开放式空间增加成员互动；整合线条让空间利落大方。

Cun–Hong Wang
Current position
Lead Designer, Wang Cun–Hong Interior Design
Summary
Skilled in giving residential spaces an expression of minimalist modernism; adjusting color contrast to increase spaciousness; design of opened spaces to improve residents' interactions; combining spatial lines to create a cleaner space.

01	02
	03 04

01 _ 走道　02 _ 主卧　03 _ 卧房　04 _ 浴室
01 _ hallway　02 _ master bedroom
03 _ bedroom　04 _ bathroom

冲突美感，让家的个性更鲜明
Conflict accentuates the character of your home

study room 书房

具冲突的情调，是本案在风格呈现上最强烈的意象。通过格局设定、层次的处理，空间的个性因而有鲜明的表现。设计师通过色彩与材质的交互运用，打造出一种都会雅痞风的居住空间，绚丽、工业感，又具有古典与现代的冲突美感。

本案的基地呈现长方形，切割为私人与公共区域，客厅、餐厅与书房因而规划为一个整体。线板通过深灰色喷漆处理，带来低调却具个性的视觉效果。ㄇ形的沙发，不但可容纳多位访客，更具有凝聚感。与客厅相比较，书房采用亮面环氧树脂以及不锈钢、玻璃等材质，展现轻盈而明亮的调性。厨房则摆上经碳化处理的废弃教室椅子，带出复古怀旧却又有创意的特色。色彩与材质在此处互相激荡，让空间更具张力。

主卧室以大图输出图案表现出强烈的视觉意象，搭配装饰性强烈的立灯、毛皮，展现私人空间奢华与舒适的一面。设计师运用多层次的设计手法，强调空间的风格，让私人宅邸的个性更加鲜明。

dining room, study room and living room 餐厅、书房与客厅

study room 书房

Photo provided_Yun-Yih Interior Design 图片提供_云邑设计

In this design, conflicting moods act as the most powerful stylistic expression. Through setting up structural layout and processing layers, the character of the space is accentuated. The designer uses a clever mix of colors and textures, to create something of a "yuppie living space". A space that is brilliant in coloration, industrial in style, but at the same time also has a sense of conflict where vintage meets modernity.

The base structure of this apartment is rectangular geometric in shape. By dividing the apartment into "private" and "public" zones, the living room, dining room and study room have become a holistic entity. The use of dark-grey spray painted line plates helps to create a visual effect that is low-key yet unique. Furthermore, the placing of a three-sided sofa not only accommodates more visitors, but also creates a sense of harmony. Compares to the living room, shiny epoxy, glasses as well as stainless steel materials are used in the study room. This creates a sense of light-weightiness and brightness in the atmosphere. As for the kitchen, a second-hand classroom chair has been re-processed through carbonation; this creates a feature that is both retro and innovative in the same time.

Colors and textures fight with each other for attention here; creating more tensions in the spatial dimensions. A strong visual image is created in the master's bedroom as it is intersected by a large image output design. This design goes well with decorative lights and furs, to display the luxurious and comfortable side to the private living space. The use of multi-dimensional design by the designer also emphasizes the character of this private apartment. A stylish statement has indeed been made in this case.

01＿主卧室　02＿主卧室　03＿餐厅、书房与客厅　04＿书房与客厅
01＿master bedroom　02＿master bedroom
03＿dining room, study room and living room
04＿study room and living room

空间案名　大来商旅
设 计 者　李中霖
设计公司　云邑室内设计
空间地点　新北市
空间性质　电梯大楼
空间面积　220 ㎡
空间格局　玄关、客厅、厨房、餐厅、书房、主卧室、主卧更衣室、主卧浴室、客房、小孩房
主要建材　玻璃、磁漆、线板

Case Name / Dai-Lai Business Hotel
Designer / Chung-Lin Li
Company / Yun-Yih Interior Design
Location / Taipei County
Nature of the apartment / Condominium
Area / $220m^2$
Outline / entrance, living room, kitchen, dining room, study room, master bedroom, master dressing room, master bedroom suite, guest room, kid's room
Main building material / glass, enamel paint, wire board

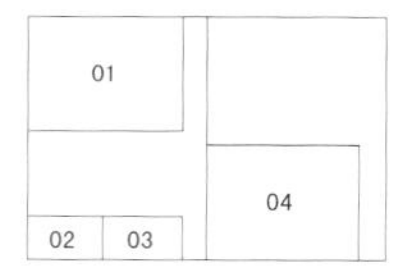

01 _ 餐厅　02 _ 餐厅与厨房　03 _ 餐厅与客厅　04 _ 书房
01 _ dining room　02 _ dining room and kitchen
03 _ dining room and living room　04 _ study room

Designer Data

李中霖
现　任/
云邑室内设计总监
学　历/
复兴美工室内设计组毕业
经　历/
1998 云邑室内设计工作室成立
2001 云邑室内设计有限公司成立
简　介/
不断地自我实践对设计的理想，总能随时与潮流接轨，以不拘泥的思维与精准落实的态度，在设计的旅程中向前迈进。

Chung-Lin Li
Current Position
Director, Yun-Yih Interior Design
Education
Fu-Hsin art school, interior design unit
Experiences
Founded Yun-Yih Interior Design Workshop in 1998
Founded Yun-Yih Interior Design, Ltd. In 2001
Brief
Always strive to realize my designing goals but in the same time catching up with the current trend. With an unconventional thinking and down to the earth attitude, I embrace my journey of interior designs ahead.

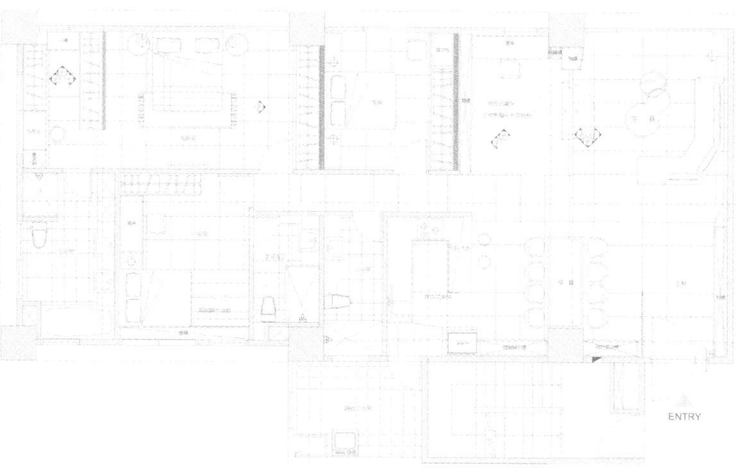

立面即空间最奢华的装饰

The most luxurious decoration is at the surface

study room 书房

living room 客厅

在这个长形的空间中，如何通过妥善的规划，让家人的凝聚力更强，是设计师思考的重点。顺应建筑结构的形式，通过将餐厅作为中介，让私人空间与公共空间完整地区隔，使格局的配置更清晰。

考虑到屋主期望父母与小孩的居住空间能相邻，既独立又不失亲密感。因此主卧室与小孩房的位置首先确立。通过设计的安排，沿着餐厅而开展的格局，正如同圆桌拉近了家人的关系，让一家人更为亲密。在本案中，空间以简洁、现代，但充满温馨的质感呈现。玄关以一面大理石作为隔屏并成为端景，其自然纹路加上自然光的点缀，在奢华中透出平淡天真的韵味。

书房空间也以大理石墙与客厅区隔，通过立面的设计，让空间各自独立，却又无阻于视野的开阔。在功能的设定上，收纳隐藏于立面的设计，将多余的线条与空间收拢干净，如电视墙与视听设备、酒柜、储藏柜均隐身于墙面之中，留下整洁的立面，供光影嬉戏、观者玩味。

What the designer wants in this case is to find a way to increase family cohesion within this long-shaped building. Conforming to the structure of the property, and to use the kitchen as an intersecting point, the private and public zones are perfectly partitioned.

Considering the fact that the house-owners want to live right next to their children out of their desire for intimacy but at the same time also want to retain their independence. For this reason, the priority is placed on establishing the locations for the master bedroom and the children' s rooms. Layout that developed along the dining room brings the family closer together just as the placing of the round table. The arrangement of the design somehow indirectly affects the relationship between the family members. In this design project, the layout is minimalistic and modern, yet full of warmth in the same time. At the entrance, a large marble stone is used as a divider. Its natural contour lines are adorned with natural lights that cast on the stone, giving the luxurious ornament a hint of pure and simplistic charm.

Marble walls are also used on the study room, which partition it with the living room. Through the careful implementation of surface design, each space is independent from each other, but in the same time causes no obstruction to the view. In turns of functionality, excessive lines and extra spaces are also hidden under the surface. For example, the TV wall, audio-visual equipment, wine cabinet as well as storage cabinets are also hidden under the wall surface, leaving a cleaner outer surface for audience to ponder; for lights to cast on.

Pictures_Ho+Hou Studio Architects Photographer_Marc Gerritsen
图片提供_何侯设计 摄影_Marc Gerritsen

House Data

空间案名 大直沈宅
设 计 者 何侯设计
参与设计 何以立、侯贞夙、钟静嘉、林于歆
空间地点 台北市大直区
空间性质 电梯大楼
空间面积 200 ㎡
空间格局 玄关、餐厅、客厅、书房、小孩房×2、客浴、主卧室、主卧更衣室、主卧浴室
主要建材 大干木木皮、橡木、璞石地板、玻璃、不锈钢、马赛克、雕刻白大理石

Case Name / Da-Tze Shen's Property
Designers / Yi-Li He, Chen-Su Hou, Ching-Chia Chung, Yu-Hsin Lin
Company / Ho+Hou Studio Architects
Location / Da-Tze District, Taipei City
Nature of the apartment/ Condominium
Area / 200m²
Outline / entrance, dining room, living room, study room, kids' room x2, guest bathroom, master bedroom, master dressing room, master bathroom
Main building material / wood veneer, oak, stone floor, glass, stainless steel, mosaic, sculpted white marble

01 _ 阳台 02 _ 餐厅与厨房 03 _ 玄关
01 _ balcony 02 _ dining room and kitchen 03 _ entrance

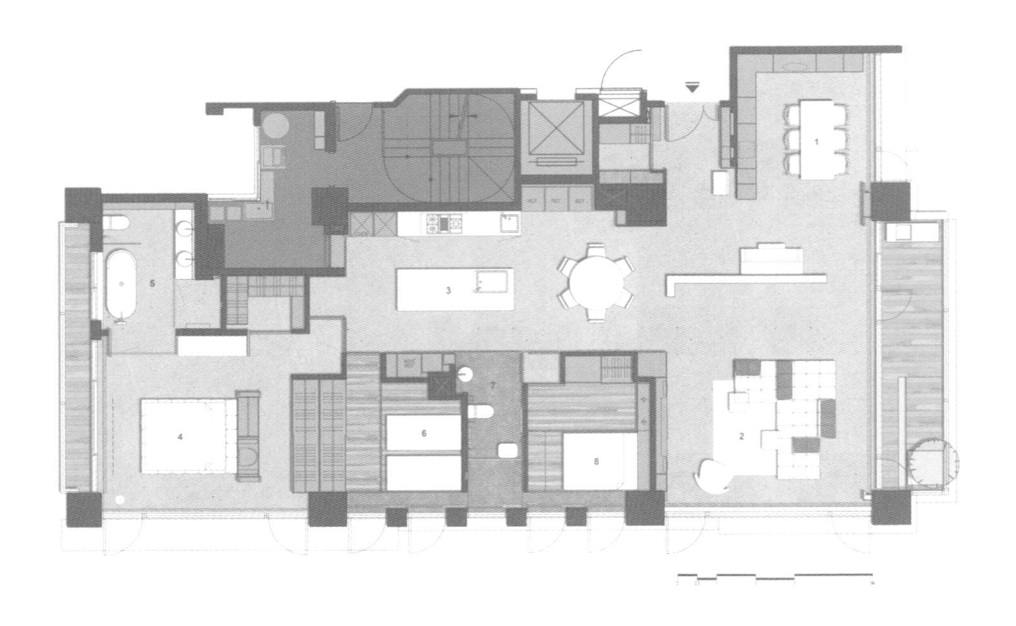

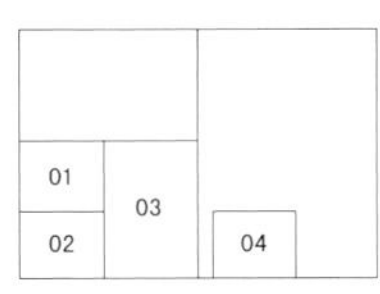

01 _ 小孩房　02 _ 客浴　03 _ 主卧浴室　04 _ 主卧室
01 _ kid's room　02 _ guest's bathroom
03 _ master bathroom　04 _ master bedroom

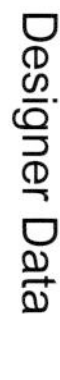

侯贞夙
现　任/
何侯设计 负责人
学　历/
1982~1985 台湾大学地理学士
1985~1988 哈佛大学景观建筑硕士
1989~1991 哥伦比亚大学建筑硕士
经　历/
1988~1989 Child Associates，波士顿，设计师。1991~1992 Rafael Vinoly Architects PC，纽约，设计师。1992~1993 Perkins & Will，纽约，设计师。1994~1997 Kohn Peterson Fox Associates PC，纽约，设计师。1996 美国纽约州注册建筑师。1997~ Ho + Hou Studio Architects，何侯设计 负责人。1997 淡江大学建筑系兼职讲师。

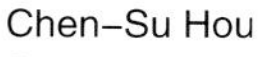

Chen-Su Hou
Current Position
Person in charge, Ho+Hou Studio Architects Education
1982-1985 Bachelor of Geography, National Taiwan University
1985-1988 Master of Landscape Architecture, Harvard University
●1989-1991 Master of Architecture, Columbia University
Experiences
●1988-1989 Designer, Child Associates, Boston,
●1991-1992 Designer , Rafael Vinoly Architects PC, New York
●1992-1993 Designer, Perkins & Will, New York
●1994-1997 Designer, Kohn Peterson Fox Associates PC,New York
●1996 Registered Architect in New York
●1997- Person in charge, Ho + Hou Studio Architects,
●1997 Adjunct Instructor of Department of Architecture, Tamkang University

何以立
现　任/
何侯设计 负责人
学　历/
1984 美国休斯敦大学建筑学士
1986 美国哈佛大学建筑系硕士
经　历/
1986 Raphael Moneo Architects，波士顿，设计师。1987~1989 Schwartz/Silver Associate Architects Inc.，波士顿，设计师。1991 美国马塞诸塞州注册建筑师。1989~1996 Pasanella + Klein Stolzman + &Berg 纽约Associate Partner，协同负责人。1997~ Ho + Hou Studio Architects，何侯设计，负责人。1996~1998 实践大学空间设计系兼职讲师。1997,2000 东海大学建筑系兼职讲师。2001~2002 台湾交通大学建筑学院兼职讲师。2003~ 敬典文教基金会执行长。 2009 台北科技大学兼职讲师。

Yi-Li He
Current Position
Person in charge, Ho+Hou Studio Architects
Education
1984 Bachelor of Architecture, University of Houston
1986 Master of Architecture, Landscape Archi-tecture, and Urban Planning, Harvard University
Experiences
●1986 Designer, Raphael Moneo Architects, Boston
●1987-2000 Designer, Schwartz/Silver Associate Architects Inc., Boston
●1991 Registered Architect in Massachusetts
●1989-1996 Pasanella + Klein Stolzman + &Berg
New York , Associate Partner
●1997- Person in charge, Ho + Hou Studio Architects
●1996-2009 Adjunct Instructor of Department of Architecture, Shih Chen University
●1997-2000 Adjunct Instructor of Department of Architecture, Tunghai University
●2001-2002 Adjunct Instructor of Graduate Institute of Architecture, Taiwan Chiao Tung University
●2003- Chief Executive Officer of HOSS Foundation
●2009 Adjunct Instructor of National Taipei University of Technology

宁静宽阔，一尘不染的都会居所

Tranquil and spotless urban accommodation

living room 客厅

Photo provided_Yun-Yih Interior Design 图片提供_云邑设计

此为18年旧屋重新改造的案例。屋子位于大厦顶楼，为复式结构。设计师考虑空间原有的结构与特性，采用区块转折的概念设计，让空间在无差别的串联中产生层次感，从而打造出融合现代感而又具冷冽气质的居所。

考虑屋主偏爱极简与现代感的设计，在风格的诠释上以洁净、内敛作为主题。纯洁的白色与深色对比，让空间展现出"而无车马喧"的独立感，仿不锈钢石英砖铺陈的地面，带来工业粗犷感，并与不锈钢材质的茶几产生对话的可能性。

客厅的挑空设计，通过垂直向度的伸展强调宽阔感，块状的设计让空间得以串联，却也能保持独立。客厅与楼梯间通过圆弧与台阶巧妙划分，餐厅则以空间格局的安排自成一区。楼梯区块几何切割的意象尤为明显，黑色烤漆玻璃、白色墙面、带状的间接灯光，呈现出材质的趣味性，并将视线层层叠叠向上延伸。二楼的私人空间以小书房为轴心，让其他空间绕其发展，展现出层次分明的格局安排。

In this case, an 18 years old house has been renovated. The apartment is located on the top floor of a building. The spatial feature the designer used is what is called "stratified design" : the designer carefully considered the structures and properties in the original space, and then using the concept of "block turning" to create a feeling of layered space. Based on this concept, a sleek and modern living environment is created.

Considering the owner's preference for minimalistic and contemporary design, the designer uses the concept of "cleanliness" and "subtleness" as the theme of the design. A sense of contrast is created by using white and dark colors together. This helps to create a feeling of space with a fare of "independence" and "calmness". Furthermore, imitation metal porcelain stoneware tiles are laid on the floors. This creates a feeling of "industrial ruggedness", which complements the coffee table made of stainless steel.

Atrium-style living room accentuates a sense of broadness through the extension of the vertical dimension. Block-like design creates continuity in space but at the same time maintains its independence. The living room and the staircase are cleverly divided through the use of arc and stage. Through the use of spatial arrangement, the dining room acts as a self-contained space. The use of geometric cuttings is especially evident at the stairway. Moreover, the use of black tinted glasses, walls painted in white, strips of indirect lighting, all create a feeling of sophistication and elegance, and at the same time provide layers and layers of vertical visual extensions. In terms of private spaces at the second floor, the small study room acts as a central point, that allows the development of other spatial arrangements to surround it. This in turn creates distinct layers in the structural arrangement of the apartment.

House Data

空间案名	北宁路
设 计 者	李中霖
参与设计	江晓慈
设计公司	云邑室内设计
空间地点	台北市
空间性质	跃层
空间面积	215 ㎡
空间格局	玄关、客厅、餐厅、厨房、休闲室、长亲房、客浴×2、主卧房、主卧浴室、起居室、小孩房×2
主要建材	仿锈铜砖、黑烤玻、磁漆，日本石

Case Name / Pei-Ning Road
Designer / Chung-Lin Li
Co-designer / Hsiao-Tzu Chiang
Company / Yun-Yih Interior Design
Location / Taipei City
Nature of the apartment / Duplex apartment
Area / 215m^2
Outline / entrance, lounge, dining room, kitchen, recreation room, elder's room, guest shower room x2, master bedroom, master bathroom, living room, kid's room x2
Main building material / copper tiles, black tinted glass, enamel paint ,Japanese-style stone

01 _ 客厅 02 _ 客厅 03 _ 客厅 04 _ 楼梯 05 _ 客厅挑空全景
01 _ living room 02 _ living room 03 _ living room
04 _ staircases 05 _ panoramic view of the living room

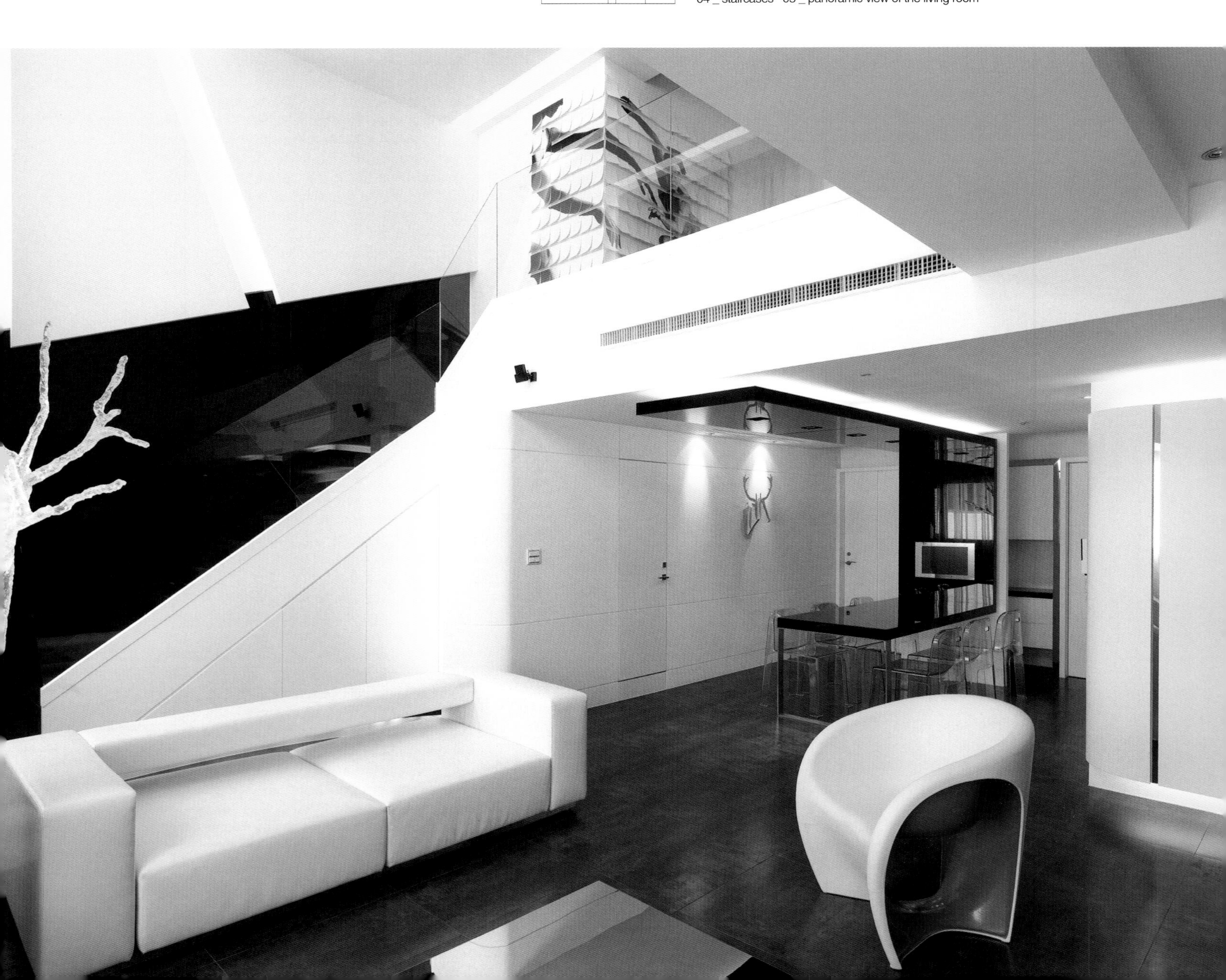

Designer Data

李中霖
现　任／
云邑室内设计总监
学　历／
复兴美工室内设计组毕业
经　历／
1998 云邑室内设计工作室成立
2001 云邑室内设计有限公司成立
简　介／
不断地自我实践对设计的理想，总能随时与潮流接轨，以不拘泥的思维与精准落实的态度，在设计的旅程中向前迈进。

Chung-Lin Li
Current Position
Director, Yun-Yih Interior Design
Education
Fu-Hsin art school, interior design unit
Experiences
Founded Yun-Yih Interior Design Workshop in 1998
Founded Yun-Yih Interior Design, Ltd. In 2001
Brief
Always strive to realize my designing goals but in the same time catching up with the current trend. With an unconventional thinking and down to the earth attitude, I embrace my journey of interior designs ahead.

01＿客厅　02＿二楼起居室　03＿餐厅与客厅
04＿客厅　05＿二楼起居室
01＿living room　02＿ living room on the second floor
03＿dining room and lounge　04＿living room
05＿living room on the second floor

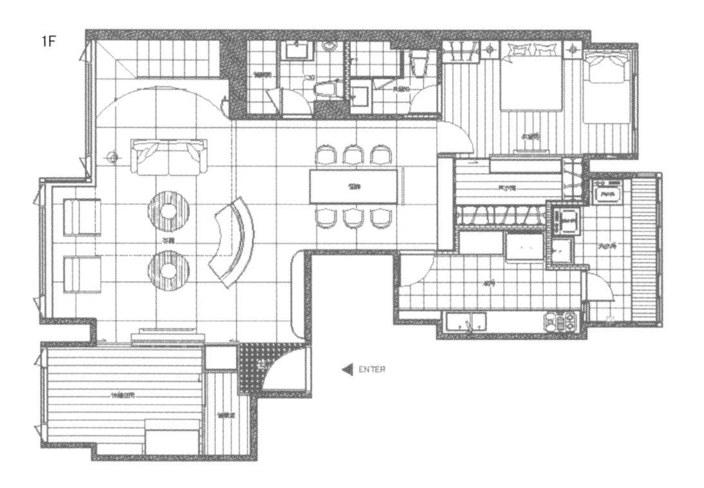

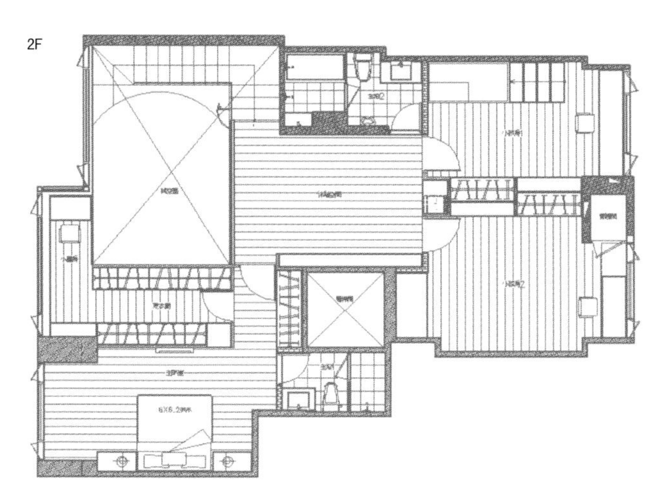

打造具呼吸感的优雅大宅

Create an elegant home that can breathe

living room 客厅

图片提供_权释国际空间设计

Pictures_Allness Design

一家3口居住在跃层空间，如何让居家空间在奢华中也显得独特？在本案中，设计师以"呼吸宅"为空间定义。借由房子原有的优势，将户外与室内的空间连接在一起，视野得以延伸，空间自然更为放大，因而营造出宽敞而没有界限的居住环境。

整个居家空间以新古典风格作为设计主调，一楼的格局最讲求的是能让空间感放大，并利用材质的搭配营造出华丽且具质感的环境。通过与庭院的连接，室内因采光的充足而带来放大的效果。大理石地砖的运用，提高了空间的质感，对称概念在此处也被细腻地融入空间，表现出新古典风格平衡稳重的一面。

由于男女主人需要各自的生活空间，因此本案采用双主卧的设计。男主人房通过英式风格中沉稳内敛的格调，表现出绅士般的优雅；女主人浪漫的个性，也由法式古典的元素表露无遗，大面积的更衣室，更能满足女主人对衣物、珠宝与名牌包的收纳需求。通过空间的定位，以及具有层次感的设计，打造出高质感而舒适的大宅。

How do you make a unique yet luxurious living space within a duplex apartment where three family members reside in? In this case, the designer uses the term "respiratory house" to define his work. By fully taking the advantage of the original outline of the house, the designer links the indoor and outdoor spaces together, so that the vision horizon can be extended. This in turn gives an illusion of a large space, creating a spacious living environment where structural boundaries are blurred.

The home itself is designed in neo–classical style. The main emphasis of the design for the layout of the first floor is to give an illusion of an enlarged space. By using a careful selection of materials an extravagant and classy living environment is created.

Furthermore, the linking of the house to the garden also brings ample amount of lighting indoor, thus making the room feels bigger. The employment of marble floor also accentuates the quality of the space, too. The concept of symmetric beauty has also been incorporated into the design, highlighting the matured and sophisticated side of neo–classical style.

Since both the host and hostess require their own private living space, two master bedrooms have been designed. The host's room is designed in a quintessential English style which is both refined and restrained. This creates an air of elegance as found in an English gentleman. The romantic nature of the hostess can also be represented by the application of yet another classical style: in this case, French. A large dressing room is built to satisfy the storage requirement for an assortment of the hostess' clothing, jewelry and designer bags. By using a design technique known as "spatial positioning", as well as using the concept of layering in design, a high quality and comfortable home environment is created.

House Data

空间案名 公园录
设 计 者 洪韡华
设计公司 权释国际空间设计
空间地点 台北
空间性质 跃层
空间面积 215 m^2
空间格局 楼下：客厅、餐厅、卫浴、男主人房、小孩房、露台。楼上：书房、女主人房、更衣室、露台
主要建材 仿鳄鱼皮、柚木、茶色镜、印度黑大理石、抛光石英砖、复古砖

Case Name / Gong Yuan Lu
Designer / Wei–Hua Hung
Company / Allness Design
Location / Taipei City
Nature of the apartment / duplex apartment
Area / 215m^2
Outline / Downstairs: living room, dining room, bathroom, host's bedroom, kid's room, terrace; upstairs: study room, hostess' room, dressing room, terrace
Main building material / alligator skin, teak, tea mirror, Indian black marble, polished porcelain tile, retro brick

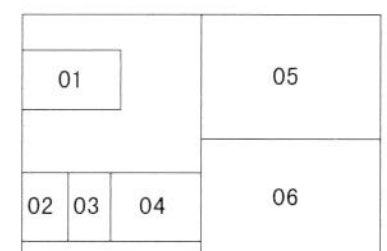

01 _ 客厅 02 _ 玄关 03 _ 楼梯 04 _ 客厅 05 _ 客厅 06 _ 客厅
01 _ living room 02 _ entrance 03 _ Stairs
04 _ living room 05 _ living room 06 _ living room

Designer Data

洪韡华
现　任/
权释国际空间设计 总监

简　介/
擅长依循屋主不同性格打造出深具东方诗意居家情境，且也能设计出最耐人寻味且百看不厌的“新大宅”样貌。著有《我家就是五星级饭店——提升居家质感设计200》。

Wei-Hua Hung
Current Position
Director, Allness Design

Brief
My specialty are to create a personalized living environment according to the personality of my client, in oriental style. I am also good at designing homes that are contemporary in style, and which you would never get bored by looking at it. I have written a book called “My home is a five-star hotel-enhance your home designing 200” .

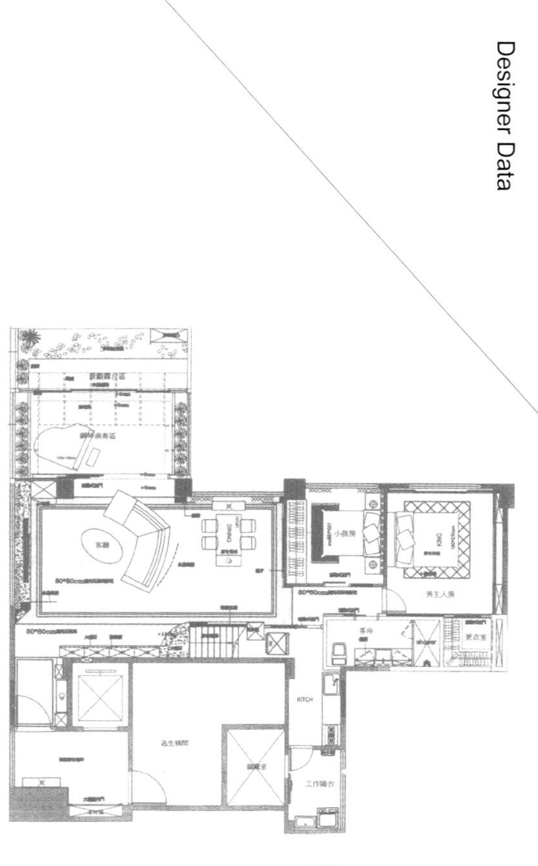

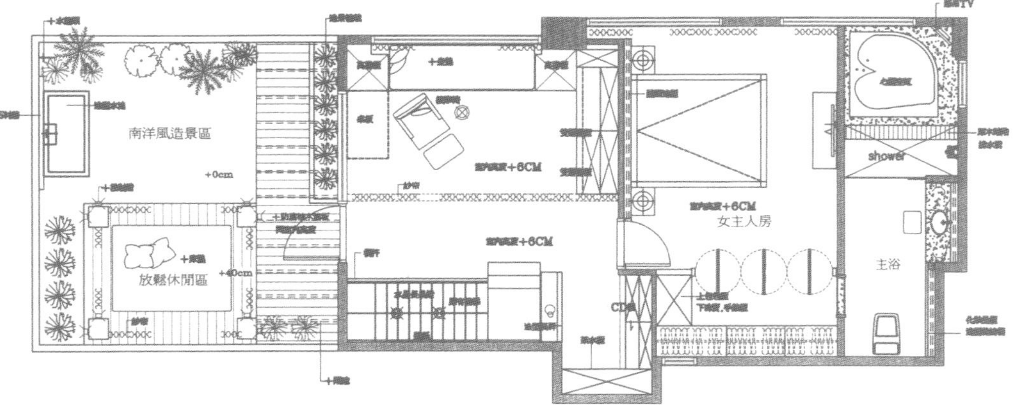

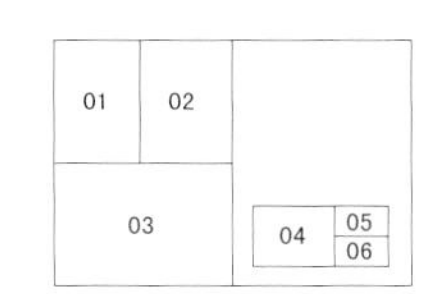

01 _ 餐厅　02 _ 女主人房浴室
03 _ 男主人房　04 _ 2楼起居室
05 _ 2楼过道　06 _ 女主人房更衣室
01 _ dining room　02 _ hostess' bathroom
03 _ host's room　04 _ 2rd floor living room
05 _ 2 F aisle　06 _ hostess' dressing room

放大，打破生活的制式框架

Enlarge and break down the rigid framework of living

本案空间呈现冂形的格局，原本中间区块有采光不足的问题，在空间的规划上也难以安排。设计师打破常规框架的设计手法，让大宅不但更显其大，而且给予屋主一个全新的居住环境。

在冂形的空间格局上，中段较为狭窄的空间作为连接两端的枢纽。结合屋主一家的生活习惯，同时也为了能通过空间的改变而拉近亲子关系，女主人专属的休憩区与琴房被配置于此处。当小孩在练琴时，母亲可以陪伴在一旁，也可避免干扰，无形中让母子的关系更为紧密。考虑到视听设备需要独立空间，男主人书房则设计了玻璃与折门，同时满足采光与区隔的功能。

公共厅区的设计融入交谊厅的概念，开放式的设计让客厅、餐厅与厨房成为一个完整的区块。通过雕塑概念的设计手法，电视柜、中岛区成为区分场域的中介，环绕式的动线刻意降低的高度放大了视野，创造了更为自由的活动区域，也展现出大宅应有的气势。

In this design case, the layout of the property is "U–shaped". There was a problem with insufficient lighting at a middle zone inside the U–shaped groove. Moreover, its awkward shape also means that architectural space planning was hard to carry out. However, the designer skillfully breaks down the rigid frames, so that not only makes the apartment looks bigger, but also provides the owner a brand new living experience.

Within the U–shaped base structure, the narrower space in the middle acts as a hub linking the two ends. This design encompasses the lifestyles of all members of the household, and at the same time makes the family come closer through spatial re–arrangement of the layout. The exclusive resting area for the hostess of the household and the piano room are both located here. When the child is playing the piano, the mother can accompany him/her in this interference–free zone. Due to this arrangement, the mother and the child are brought closer together. As for the study room for the male host, the designer has considered the fact that his audio–visual equipment requires a private and enclosed space to play. For this reason, glasses and folding door are used in his design, from which the purpose of room partitioning as well as providing sufficient lighting are both fulfilled.

The concept of common room has been integrated into the design of the public area. The employment of open design allows the formation of a complete block encompassing the living room, dining room and kitchen. Through the use of sculpturing design techniques, the TV cabinet, and the "mid–island" has become boundary lines that intersect different zones. The use of encircling routes also allows more freedom of movement. Deliberately lowered ceiling also creates an illusion of broadened horizon, as well as making the apartment more imposing.

dining room 餐厅

Pictures_ Materiality Design Photographer_ Kyle Yu
图片提供_石坊空间设计研究事务所 摄影_Kyle Yu

House Data

空间案名 舒放生活
设 计 者 郭宗翰
参与设计 陈婷亮
设计公司 石坊空间设计研究
空间地点 台北市
空间性质 电梯大楼
空间面积 202 m^2
空间格局 客厅、餐厅、厨房、女主人开放休憩区、男主人书房、主卧室、主浴室、客浴室、小孩房、练琴区
主要建材 石材、玻璃、实木地板、烤漆木作、实木、进口家饰布、铝件、柚木

Case Name / A relaxed life
Designer / Tsung-Han Kuo
Co-designer / Ting-Liang Chen
Company / Materiality Design
Location / Taipei City
Nature of the apartment / Condominium
Area / $202m^2$
Outline / living room, dining room, kitchen, open resting area for the hostess, the host's study room, master bedroom, master bathroom, guest bathroom, kid's room, piano practicing area
Main building material / stone, glass, hardwood floor, laminated wood, hardwood, imported furniture fabrics, aluminum, teak-wood

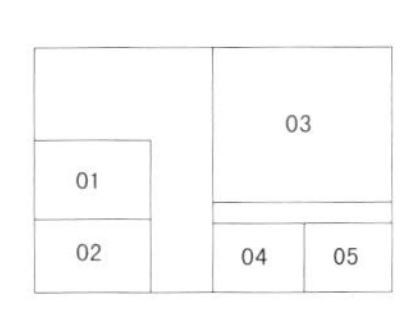

01 _ 客厅与餐厅 02 _ 客厅 03 _ 男主人书房
04 _ 女主人书房 05 _ 女主人书房与琴房
01 _ living room and dining room 02 _ living room
03 _ host's study room 04 _ hostess' study room
05 _ hostess' study room and piano room

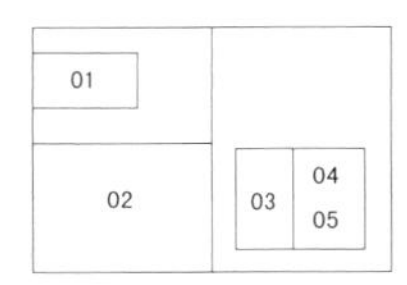

01 _ 主卧浴室　02 _ 男主人书房与过道　03 _ 小孩房　04 _ 主卧浴室　05 _ 主卧房
01 _ master bathroom　02 _ host's study room and the aisle
03 _ kid's room　04 _ master bathroom　05 _ master bathroom

Designer Data

郭宗翰
现　任／
石坊空间设计研究　设计总监
学　历／
1997~1999　英国伦敦艺术大学空间设计系学士
1999~2000　英国北伦敦大学建筑设计系硕士
经　历／
2000~2002　香港商穆氏设计　设计师
2002　石坊空间设计研究　设计总监
2005　实践大学设计学院兼职讲师

Tsung-Han Kuo
Current Position
Director, Materiality Design
Education
1997-1999　University of the Arts London, BA in Spatial Design
1999-2000　University of North London, UK , MA in architectural design
Experiences
2000-2002　Designer , Moody's Commercial Design, Hong Kong
2002　Director, Materiality Design
2005　Lecturer at College of Design, Shih Chien University

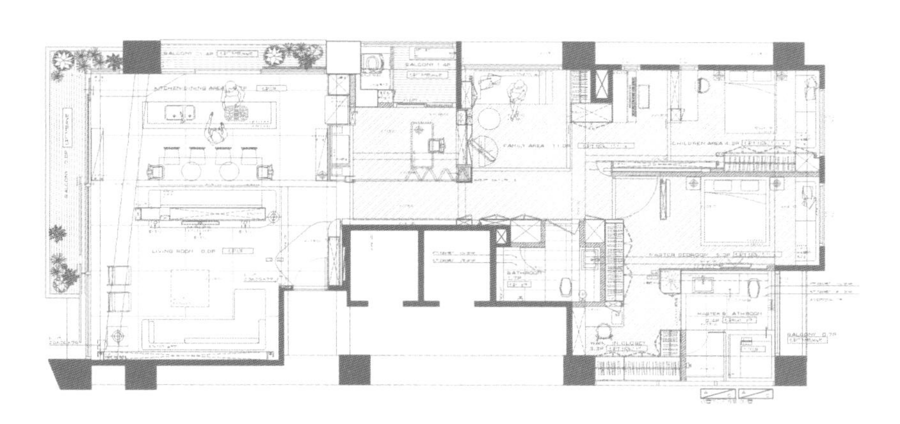

纯化空间，奢华源自片刻的宁静

Purify the space: luxury is found in the moment of tranquility

Pictures_Atelier Construction de l'Espace Photographer_ Chi-Min Wu
图片提供_空间制作所 摄影_吴启民

追求"感知"上的平衡和谐，以及如何通过空间设计创造一个宁静的生活场所，是本案设计的一个关键。结合空间三面采光的优势，设计师需要做的，便是将空间做最妥帖的规划，借助空间的精准布局，让居住场所能有纯净的表现。

由于基地为长形，为避免出现走廊，设计师在空间中设计了一个"盒子"。"盒子"既是储藏室、准备室，同时又是整个空间的一个枢纽。"盒子"的存在，为玄关创造出两条动线，也正好界定了客厅的区域，并且替三代同堂的居住者创造理想的生活方式。玄关进入客厅的行进路线，受到客厅自然光线的引导而营造出宽阔感。长亲房与主卧室、小孩房正好配置于房子的两侧，但又能因公共空间的设定而拉近彼此的距离。

讲求"宽松"与"单纯"的氛围，设计师在此处使用低彩度的色彩设计，利用仿石材砖、灰紫蓝色墙面铺陈出空间的宁静感。而天花板使用皮革材质，以倒"L"形框框出电视墙，既展现低调奢华的质感，又为空间创造出视觉重心。借助细腻的手法与安排，空间层次如水墨画般展开，让每一个空间都能感受到独特的宁静。

The main concept of this design is to create a tranquil living environment where there is a sense of balance in the perception of the inhabitants. Since the building has an advantage of being well-lit from three light sources, what the designer needs to do is to incorporate these advantages into the design, allowing a more appropriate spatial layout planning. Through this precise implementation of spatial layout design, the living environment has become "purer" and more clear-cut.

Since the foundation of the apartment is long-shaped, in order to avoid the existence of a corridor, the designer has designed a "box" at the middle part of the apartment. This box acts as a storage room, a preparatory room, as well as playing a pivotal role of the entire living space. The presence of this box helps to create a second route at the entrance, and at the same time defining a zone for the living room, as well as creating an ideal living environment for three-generations of family that lives here. Through the effect of natural lighting, the route which enters the living room gives an illusion of a larger space. Although the elderly' s room, the master bedroom and the kid' s room are located at the two sides of the apartment, the inhabitants are made to come closer together through the provision of the public space in between.

The emphasis of this design is to create an atmosphere which is "relaxed" and "pure". For this reason, the designer uses materials which are low in saturation: the use of imitated stone tiles as well as the wall painted in purplish-blue all helps to create a sense of tranquility. The ceiling uses leathered materials, which helps to define the TV wall through its reverse L-shaped frame. Not only does it create a sense of toned-downed luxury but also create a visual point to the design. Through this intricate design and planning, the sense of layers is produced, allowing each space exhibiting a sense of serenity.

House Data

空间案名　陈宅
设 计 者　高弘树
参与设计　陈薇如、郑贵凤
设计公司　制作空间室内装修设计工程有限公司
空间地点　台北市
空间性质　电梯大楼
空间面积　198 ㎡
空间格局　玄关、客厅、厨房、餐厅、主卧室、主卧浴室、主卧更衣室、小孩房、储藏室
主要建材　意大利仿鹅卵石石英砖、染深灰橡木、橡木集成地板、乳胶漆、人造皮革

Case Name / Chen's Property
Designer / Hung-Shu Kao
Co-designer / Wei-Ju Chen, Kui-Feng Cheng
Company / Atelier Construction de l'Espace
Location / Taipei City
Nature of the apartment / Condominium
Area / 198 m^2
Outline / entrance, living room, kitchen, dining room, master bedroom, master bathroom, master dressing room, kid's room, storage
Main building material / Italian imitated-cobblestone porcelain tile, dark grey-dyed oak, oak composite floor, ICI computerized coloring latex paint, synthetic leather

01 02 03 04

01 _ 客厅　02 _ 客厅　03 _ 玄关　04 _ 餐厅
01 _ living room　02 _ living room　03 _ entrance　04 _ dining room

Designer Data

高弘树
现　任／
空间制作所Atelier Construction de l'Espace
主持设计师
制作空间室内装修设计工程有限公司负责人
学　历／
1989 逢甲大学建筑系工学士
1995 法国建筑专业学院建筑师文凭（DESA, Diplôme d'Architecte de l'Ecole Spéciale d'Architecture）
经　历／
1995～1997 沈祖海建筑师事务所／项目设计师
1998～2000 私立实践大学室内空间设计系兼职讲师
1999～2005 台北科技大学建筑系兼职讲师
2006～　学学文创志业空间研究室讲师
2006～　台北科技大学设计学院兼职讲师

Hung-Shu Kao
Current Position
Atelier Construction de l'Espace Chief Designer
Person in charge, Atelier Construction de l'Espace
Education
1989　Bachelor of Architecture, Feng Chia University
1995　DESA, Diplôme d'Architecte de l'Ecole Spéciale d'Architecture
Experiences
1995–1997 Project Designer, Shen-Zu-Hai Architectural Design Firm
1998–2000 Adjunct Instructor of Department of Architecture, Shih Chien University
1999–2005 Adjunct Instructor of Department of Architecture, Taipei University of Technology
2006– Instructor of Interior Design Workshop, Xue Xue Institute
2006– Adjunct Instructor of College of Design, Taipei University of Technology

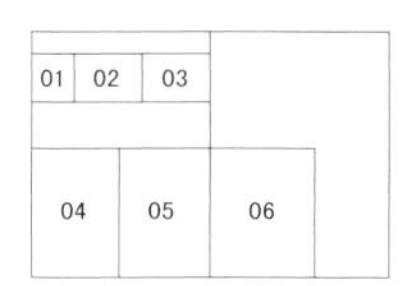

01 _ 主卧室　02 _ 主卧浴室　03 _ 玄关过道与主卧起居室
04 _ 餐厅　05 _ 主卧浴室　06 _ 玄关过道
01 _ master bedroom　02 _ master bathroom
03 _ entrance hallway and living room, master bedroom
04 _ dining room　05 _ master bathroom　06 _ entrance corridor

hallway 玄关

艺术与质感成就奢华居家

Art and class lead to a luxurious house

living room 客厅

living room 客厅

质感的创造来自于艺术品位的延伸。在这间私人住宅里，设计师掌握住空间的尺度，通过具体的规划，将屋主对于居住的需求融于设计细节里。而空间气质与深度的营造，则交由艺术品本身发挥，让家成为一个微型博物馆。

空间的宽敞感来自于公共空间的定位，从玄关入口首先可见宽敞的客厅，方正的格局为营造一个可放松也可接待访客的空间。大面积的茶色镜则从客厅转折至餐厅空间，空间的串联、镜面的反射，加之视线的引导效果，让公共空间的向度加倍放大。书房与客厅间则利用可穿透的展示柜区隔，隐藏式灯光成为发亮的画框，由艺术品演绎的设计语汇让空间感饶富趣味。

开放式的空间在一定程度上仍需区隔，故在厨房与餐厅间使用拉门隔断，让空间的使用更灵活。私人空间设计讲求收纳功能的满足，以及舒适感营造，通过床头背墙的质感提升，睡眠区域也因此拥有媲美饭店的质感。本案通过亨利摩尔、安迪沃荷、常玉以及吴冠中等人的作品，诉说屋主的品位，也将艺术美感融合于空间。

Class comes from the extension of art taste. The designer fulfills the owner's needs in detail by grasping the space and planning the layout of the house. Moreover, the artworks will create the class and depth for the house, making a house a micro museum.

The spaciousness came from the location of the public area. Entering the house, you will see an airy living room. The square layout creates a space for relaxing and hosting the guests. A large bronze mirror extends from the living room to a dining room, guiding the route, connecting the spaces, and magnifying the public area. The study and the living room are divided by the transparent showcase. The hidden lighting serves as the shining picture frames. The design language of artworks makes the space very interesting.

The open area still needs divisions. Therefore, the kitchen and the dining room are divided by a sliding door, which makes the use of the space more flexible. The private area focuses on storage functions and coziness. The sleep area is as good as hotels in terms of quietness and quality as the head of the bed is against the wall. The artworks of Henry Moore, Andy Warhol, Yu Chang and Guan-zhong Wu tell of the dwellers' taste and integrate the sense of art to the space.

Pictures_TAD ASSOCIATES INC
图片提供_大隐设计

01 _ 客厅　02 _ 客厅与书房　03 _ 客厅　04 _ 餐厅与厨房　05 _ 餐厅
01 _ living room　02 _ living room and study room　03 _ living room
04 _ dinning room and kitchen　05 _ dinning room

空间案名　南海苑
设 计 者　王镇华
设计公司　大隐设计集团
空间地点　台北市
空间性质　电梯大楼
空间面积　190 m²
空间格局　玄关、客厅、餐厅、厨房、书房、主卧室、主卧浴室、主卧更衣室、客房×2、客浴×2
主要建材　喷漆、茶色镜、柚木地板、染白白橡木木皮、皮革

Title / South Sea Mansion
Designer / Arthur Wang
Design company / TAD ASSOCIATES INC.
Location / Taipei City
Category / Elevator building
Area / 190 m²
Layout / Entrance, living room, dining room, kitchen, study, master bedroom, master bathroom, master dressing room, guest room x2, guest bathroom x2
Building materials / tea-colored mirrors, teak flooring, dyed white cherry bark, leather, imported wallpapers

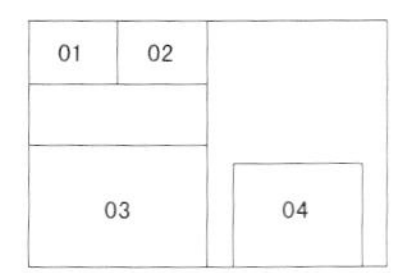

01 _ 小孩房　02 _ 主卧室　03 _ 主卧室与更衣室　04 _ 主卧浴室
01 _ child room　02 _ master bedroom
03 _ master bedroom and changing room　04 _ master bathroom

Designer Data

王镇华

现　任/

大隐设计集团台北、上海主持人

学　历/

辅仁大学毕业

经　历/

1985~1989　珠宝设计师

1999　罗芙奥拍卖公司、睿芙奥艺术顾问公司

简　介/

王镇华从事室内设计超过20年，在这期间陆续成立了A&F设计事务所、大隐室内设计公司(TAD ASSOCIATES INC.)，并将事业领域扩展至中国内地。王镇华以其对艺术品长期涉猎的经验，为委托客户提供艺术品陈设及收藏的顾问。目前其所主持的设计公司主要以承接私人豪宅以及企业招待所、办公空间为主。

Arthur Wang

Principal of TAD ASSOCIATES INC., Taipei and TAD ASSOCIATES INC., Shanghai

Professional

1985–1989　Jewelry designer of ALL GEMS JEWELRY

1999　Ravenel Art Group

Education

Fu Jen Catholic University

Brief bio

Arthur Wang has been in the interior design for more than twenty years. He founded A&F INC and TAD ASSOCIATES INC and has expanded his business to China. Wang is an artworks counselor for his clients. His companies focus on the interior design service for private mansions, enterprise guest houses and offices.

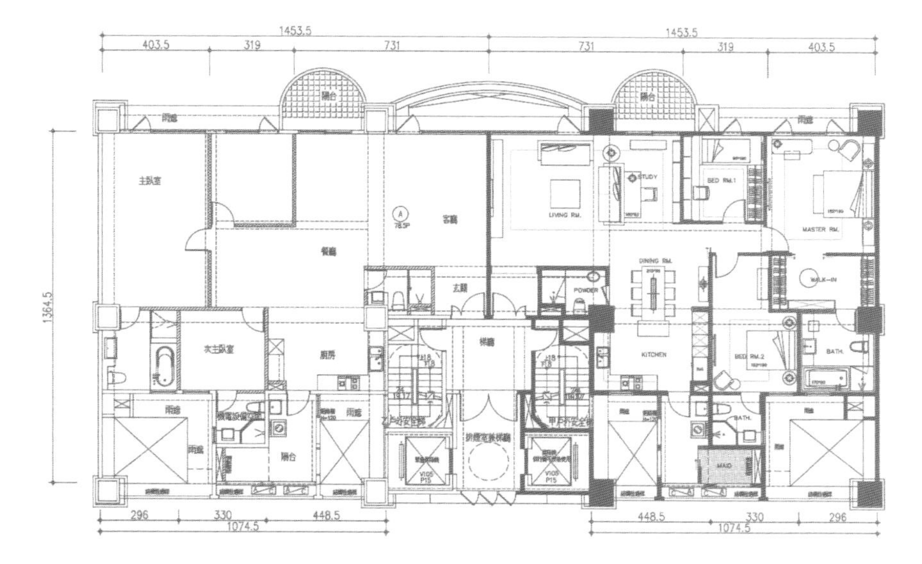

hallway and living room玄关与客厅

时尚与休闲的交会点

Intersection between fashion and leisure

living room and dining room客厅与餐厅

弧线以及自由度，是这个空间最重要的两个设计语汇。从客餐厅的串联开始，空间受到弧线的引导，一路从客厅、餐厅延伸到了书房以及主卧室。设计师在这个案子中，将“流畅”的意涵发挥得淋漓尽致，打造出清爽而又具朝气的大宅。

从大门进入，可以看到大面积的窗户以及沿墙而出的矮柜，电视正好从中间将客厅与餐厅区隔成两个区块。矮柜的弧形将视线带到书房与餐厅的隔墙，弧线至此垂直面折进了走道，暗示出私人空间的位置。天花板的弧形是地面弧形的呼应，同时也呈现出雕塑般的立体感。由于书房正好邻接餐厅，因此通过以倒梯形的玻璃作为隔断，既产生视觉通透的效果，也让公共区域的视线得到延伸。主卧的设计以衣柜作为床头墙，并创造出了环绕式的动线。浴室的入口同时也可通往更衣室。

弧线的设计语汇让空间更显自由，而家具的选配以及色系上的铺陈，都强调出了居住者的年轻背景，让家成为休闲而时尚的场所。

Curve lines and degree of freedom are the most important concepts of this space. The place is guided by a curve line, which extends along the living room, dining room, study and the master bedroom. The designer of the project, Shi-chieh Lu, exploits the idea of "smoothness" by creating a fresh and energetic mansion.

Entering from the gate, you will see a big window and the credenza that seems to grow from the wall. The living room and the dining room are divided by the television, right down the middle. The arc of the credenza instructs us to see the partition wall of the study and the dining room, the curve is hidden in the hallway, implying that there is a private space inside. The arc of the ceiling echoes with the arc on the ground, and also presents the sculpture-like atmosphere. The study is right next to the dining room; therefore an upside down trapezoid glass partition allows people to see beyond the room and extend the visibility range of the public area.

In the master bedroom, the designer uses a closet as the wall bed and creates a circular route. The entrance of the shower room is also the passage to the dressing room. The theme of arc and curved surface breathe into the air of freedom to the space. Moreover, the selection and allocation of the furniture and color arrangement makes the home become an exhibition area, highlighting the habitant's youth hood.

图片提供_陆希杰设计事业有限公司　摄影_Marc Gerritsen

Pictures_C J Studio Photographer_ Marc Gerritsen

living room 客厅

House Data

空间案名　大安花园王宅
设 计 者　陆希杰
参与设计　曹车政、郑家皓、刘冠汉
空间地点　台北
空间性质　单层住宅
空间面积　189 m^2
空间格局　客厅、餐厅、厨房、书房、主卧室、主卧浴室、客房、客浴
主要建材　盘多磨、美耐板、集成材木皮

Title / Wang's house, Da-an Garden
Designer / Shi-chieh Lu
Assistant designer / Che-zheng Cao, Jia-hao Zheng, Guan-han Liu
Location / Taipei City
Category / Ground floor apartment
Area / 189 m^2
Layout / Living room, dining room, kitchen, study, master bedroom, master bathroom, guest room, and guest bathroom
Building material / Pandomo, melamine board, laminated veneer

01 02	04
03	05

01 _ 客厅、餐厅及厨房　02 _ 书房　03 _ 客厅与餐厅　04 _ 客厅与餐厅　05 _ 书房
01 _ living room and dinning room and Kitchen　02 _ study room　03 _ living room and dinning room　04 _ living room and dinning room　05 _ study room

01 _ 主卧室　02 _ 主卧浴室　03 _ 主卧室　04 _ 主卧室　05 _ 主卧室电视墙
01 _ master bedroom　02 _ master bathroom　03 _ master bedroom
04 _ master bedroom　05 _ TV-wall of the master bedroom

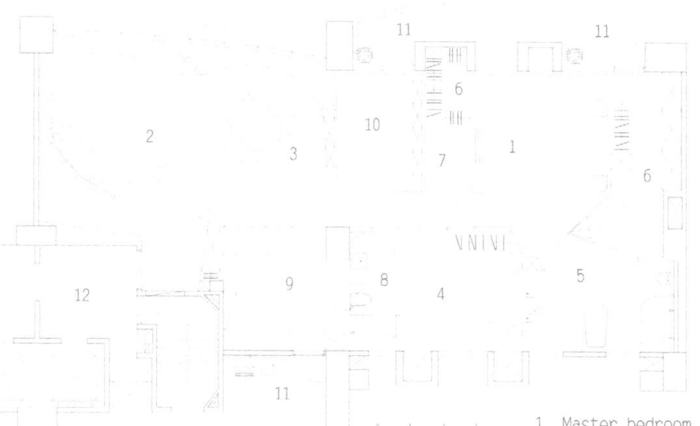

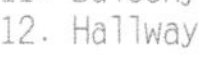

Designer Data

陆希杰
现　任／
CJ STUDIO 主持人
学　历／
东海大学建筑系毕业。英国AA建筑联盟硕士
简　介／
从事建筑及室内设计、家具设计、产品设计以及相关领域的研发，现兼任交通大学建筑研究所助理教授，著有《锻造视差》。代表作品有："窦腾璜及张李玉菁wum 概念店"；"窦腾璜及张李玉菁台中Tiger city概念店"（2005年获得日本JCD设计大赏）；"Aesop微风店"（2006年获得日本JCD设计大奖、国际室内设计联盟IFI 2007设计金奖）；"敦南国美蔡宅作品"（获得台湾室内设计大奖TID Award 2007金奖）。入选美国室内设计杂志《INTERIOR DESIGN七十五周年特集》中五位具潜力设计师之一。2008年，参展"第十一届威尼斯建筑双年展"和台北当代艺术馆"第六届城市行动艺术节"。

Shi-chieh Lu
Professional
CJ STUDIO, design director
Education
Master of Architectural Association School of Architecture, UK
Department of Architecture, Tunghai University
Brief Bio
Shichieh Lu is an adjunct assistant professor in Graduate Institute of Architecture, NTCU, specializing in interior design, furniture design and product design. He authors
Forging parallax, published in 2003. Lu's masterpieces include
• Stephanie Dou & Changlee Yugin's concept store, WUM
• Stephan Dou & Changlee Yugin's concept store in Taichung, Tiger City (won 2005 JCD Design Award)
• Aesop in Breeze Store (won 2006 JCD design award and IFI 2007 golden prize)
• Cai's house in Dun-nan Guo-mei (won TID Award 2007 golden prize) .
Lu, listed as one of the five promising architects in Interior Design Magazine's special issue for its 75 anniversary, participated in the 2008 Biennale Architecture 11th International and 2008 MOCA TAIPEI-dark city

优雅如画的生活场所

A living environment that is as graceful as a painting

dining room 餐厅

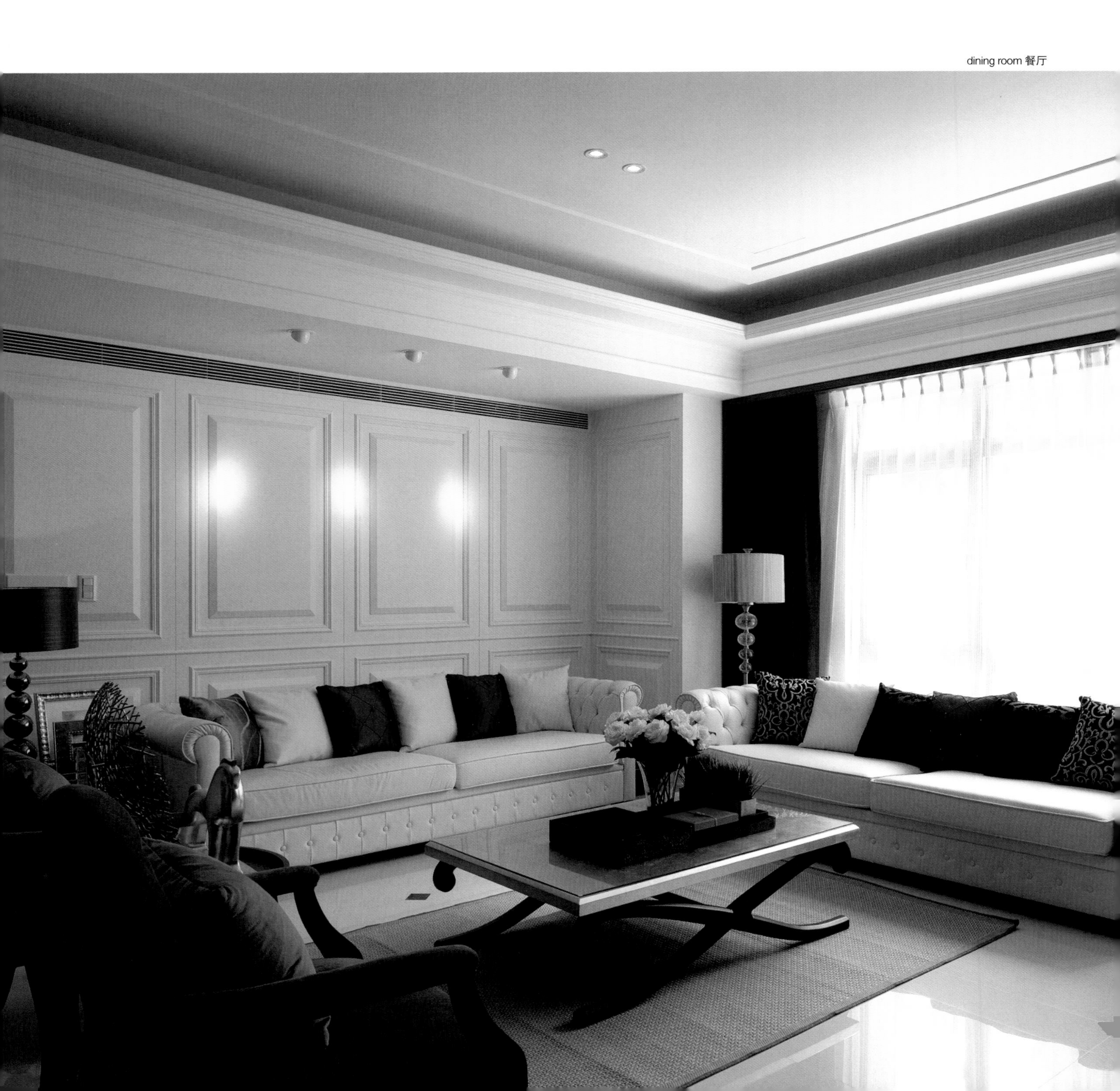

dining room and living room 餐厅与客厅

居住在大都会中，总希望能够闹中取静，而要达到真正的宁静，则可通过设计来实现。屋主夫妻期望居住的场所能传达出他们对于生活质量的要求，并且创造出属于个人的风格。因此，在本案中，通过结合空间的采光条件及配合设计巧思，替屋主打造出如画的生活空间。

公共空间的设计以能传达出优雅、舒适的风格为主。因此地面运用石材滚边的方式，铺陈出小而精致的滚边与拼花图案，而开放式的空间设计，让客厅的自然光线可进入室内，让空间更显清亮通透。天花板的造型设计整合了空间格局，创造出放大的视觉效果，因而让空间更显宽敞。至于空间的界定，则运用巧妙的手法，如走道与餐厅间运用金属珠帘区隔，解决了格局上的难题。对称式的设计与线板的运用，以及使用镜面与丝质布裱褙墙面的细腻手法，更创造出空间的奢华与不凡。

考虑到男女主人各有不同的生活习惯，分别规划了两个主卧室。女主人卧室以高雅而浪漫的风格为主。男主人卧室则可穿越进入小花园，借由户内外空间的结合，创造出与自然共生的生活意象。

One always strives for a quiet location to reside when you are living at a large city. However, in order to achieve true tranquility in life, it has to be done through interior designing. The homeowner couple expects the design to reflect their need for a good quality of living and in the same time creating a something of a personalized taste. Therefore, in this case, both lighting conditions as well as originality in design are taken into account, in order to create a living space that is as beautiful as a painting for the house owners.

In terms of the design for the public zones, main emphasis lays in delivering an elegant and comfortable style. For this reason, stone floor piping is used, which helps to create delicate embroidered borders and mosaic patterns. As for the living room, an open space design is used so that natural lights can come into the room, making the space brighter and more transparent. The ceiling design dominates the spatial layout, creating a magnified visual effect, making the room feeling more spacious. As for the partitioning of the space, a clever approach is used: metal bead curtain is used as a divider between the aisle and the dining room. This in turn solves a problem frequently encountered in spatial layout design. Moreover, a luxurious and extraordinary space is generated by the employment of symmetrical design in addition to the use of line boards, as well as using mirrors and silk clothes to decorate the walls

In terms of the bedroom design, considering the fact that both the host and the hostess of the house have different living habits, two master rooms are designed. The room for the hostess has an atmosphere of elegance and romance. As for the room for the host, it is connected to a small garden. Through this combination of the indoor and outdoor space, a room that shares a "symbiotic relationship" with nature is created.

Pictures_ Interplay Space Design Studio
图片提供_演拓空间室内设计

空间案名　神奇山庄王宅
设 计 者　张德良、殷崇渊
参与设计　演拓空间室内设计
空间地点　台北市
空间性质　电梯大楼
空间面积　167 m²
空间格局　玄关、客厅、厨房、书房、客房、客浴、主卧室×2、主卧浴室、主卧更衣室、庭院
主要建材　粗面处理大理石、抛光石英砖、墨镜、壁纸、画框、木地板

Case Name / Amazing mountain villa wangzhai
Designers / Te–Liang Chang, Chung–Yuen Yin
Company / Interplay Space Design Studio
Location / Taipei City
Nature of the apartment/ Condominium
Area / 167 m²
Outline / entrance, living room, kitchen, study room, guest's room, guest's bathroom, master bedroom x2, master bathroom, master dressing room, garden
Main building material / marble with rough surface handling, polished porcelain tile, sunglasses, wallpaper, picture frames, wooden floor

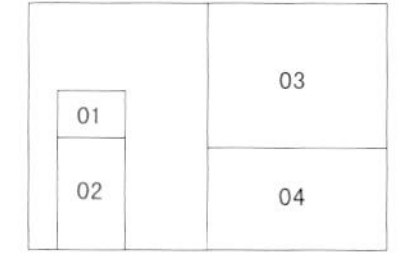

01 _ 餐厅　02 _ 玄关　03 _ 餐厅与客厅　04 _ 餐厅
01 _ dining room　02 _ entrance
03 _ dining room and living room
04 _ dinning room

Designer Data

张德良、殷崇渊
现　任/
演拓空间室内设计 总监
经　历/
台北市室内设计装修商业同业公会会员
1999 创立演拓空间室内设计
简　介/
坚持好的设计应是体贴使用的人。设计的目的在于使生活简单，而不是让居住者花过多的时间整理照顾，不应该让不合宜的设计造成居住者的负担。

Te–Liang Chang, Chung–Yuen Yin
Current Position
Director, Interplay Space Design Studio
Experiences
Member of Taipei Association of Interior Designers
1999– Established Interplay Space Design Studio
Brief
A good designer should be considerate towards his/her client. The design aims to make life simpler, rather than letting the residents spending too much time taking care of their property; and should not cause burden for the inhabitants with inappropriate designs.

01 _ 书房　02 _ 女主人房　03 _ 男主人房　04 _ 男主人房
01 _ study room　02 _ hostess' room
03 _ host's room　04 _ host's room

家的每一面都是街道风景

Each side of the house is the street view

living room 客厅

hallway 玄关

dining room 餐厅

位于都会空间的居所应该是什么样子？设计师从纽约一条充满文化艺术、深具活力的“切尔西街”寻找到灵感，而那富含表情的街道风景，便是这间房子的精神。

为了整合原空间中散落的四根大立柱，设计师利用原有的建筑结构，切割出玄关、公共区域以及私人区域，并以连续的立面设计隐藏梁柱。而随着空间的展开，设计师也将街道风景的概念融入功能之中。比如电视墙是贩卖旧书古物的二手书店，收藏背包客屋主旅行的纪念品。而白色建筑的分割线隐藏入口，其实是通往卧室的走道。看似各异的风景，借由水平线切割的延伸达成一致性，形成完整的面貌。

除了通过连续性的立面让各空间具有关联性外，在许多地方细腻设计也都蕴含无限巧思。比如客厅经过平面化处理的一盏立灯隐藏在白墙上，白天看起来相当低调，但晚上绽放光芒却又创造丰富的表情；卧室床头柜刻意留出的2cm缝隙，也为虚空间带来更多想象。温暖而又具有活力，是设计师赋予空间的调性，也呼应了背包客屋主的旅行记忆。

What is a living place in an urban area supposed to look like? The designer was inspired by Chelsea Street, full of art and energy. And the house's spirit is street views of the eclectic expressions.

To integrate its four pillars scattered in the space before renovation, the designer uses its previous architecture structure to segment entrance, public and private areas. The designer also applies the concept of fa-ade design into the living functions. For example, TV wall serves as the secondhand bookstore's shelves and accommodates the souvenirs of the owner, a backpacker. And the hidden entrance of the white building's area separation line is actually the hallway to the bedroom. The horizontal separation line unites the seemingly different views and presents consistency and completeness.

The designer carefully elaborates his ingenuity everywhere in addition to uniting the space with successive facades to enhance the connection of every space. For example, a floor lamp is hidden in the white wall, looks low-key in the day. But when it glows at night, the light creates rich expression. Or the bedside cupboard deliberately leaves 2cm space to develop more imagination for the unoccupied space. The designer endows the space with warmth and vitality, which also echoes the owner's backpacking travel memories.

图片提供_水相设计
Pictures_Waterfrom Design

空间案名　忠孝钟宅
设 计 者　李智翔
设计公司　水相设计
空间地点　台北市
空间性质　电梯大楼
空间面积　166.5 m^2
空间格局　客厅、厨房
主要建材　染白栓木、烤漆、风化木、欧松板、帝诺大理石

Title / Chelsea street
Designer / Zhi-xiang Li
Design company / Waterfrom Design
Location / Taipei City
Category / Elevator building
Area / 166.5 m^2
Layout / Living room and kitchen
Building material / White doughwood, baking varnish, weathered wood, OSB board, and Diano marble slab

01 _ 客厅　02 _ 客厅电视墙　03 _ 餐厅　04 _ 客厅
05 _ 餐厅与客厅
01 _ living room　02 _ TV-wall
03 _ dining room　04 _ living room
05 _ living room and dining room

01 _ 客厅与餐厅　02 _ 餐厅　03 _ 书房望向客厅　04 _ 主卧　05 _ 客厅　06 _ 休憩室　07 _ 主卧、主浴
01 _ living room and dining room　02 _ dining room　03 _ study towards the living room
04 _ master bedroom　05 _ drawing room　06 _ master bedroom and master bathroom

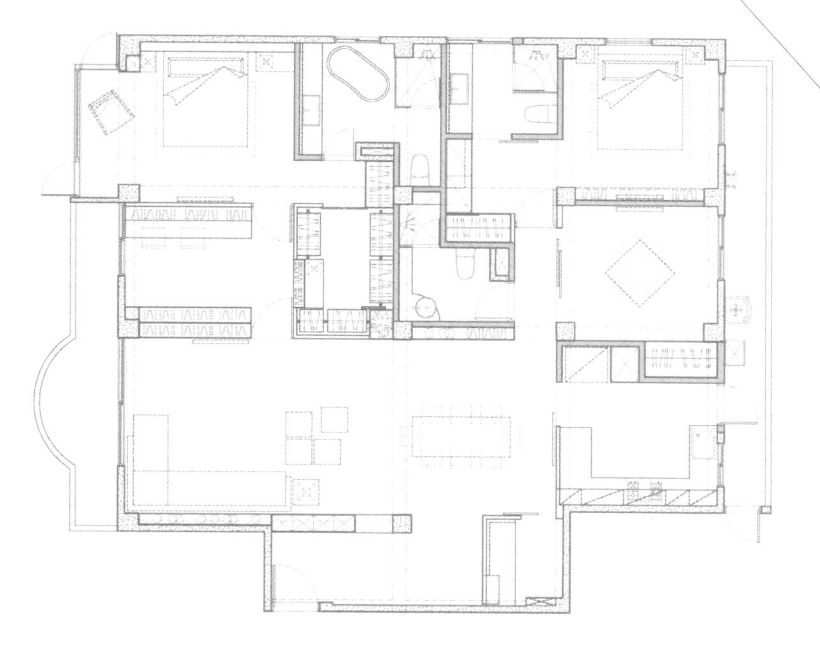

李智翔

现　任／

水相设计　设计总监

学　历／

1998 ~ 1999　丹麦哥本哈根大学研修毕业

1997 ~ 2000　美国纽约布拉特学院(Pratt Institute)室内设计硕士

经　历／

2008　水相设计 设计总监

2004 ~ 2008　荷果设计 设计总监

2001 ~ 2003　MMoser 香港商穆氏设计集团

2000 ~ 2001　李玮珉建筑师事务所

简　介／

重新再定义空间，创造真正量身订制的个性住宅，富含幽默感的设计语汇，是李智翔设计师不断尝试在空间注入的特质。近期作品更追求“自然光与线条比例”的相对关系，以更内敛但不含蓄的手法处理每一项细节，创造出独特的设计风格。

Zhi-xiang Li

Waterfrom Design, Design director

Education

1998-1999　The University of Copenhagen, Research programmes

1997-2000　Pratt Institute, New York,USA. Mater of Interior Design

Professional

2008-present　Waterfrom Design, Design director,

2004-2008　Hercore Design, Design director,

2001-2003　MMoser

2000-2001　LWMA

Brief bio

Li Zhi-xiang attempts to instill the space with the refreshing quality by redefining the space, creating the tailor-made personal houses, and employing humorous vocabulary. His recent works pursue the relative relationship between "natural light and linear proportion." And Li handles every detail in a restrained but not reserved way, in order to create absolutely unique designs.

优雅时尚，大宅新定义

Elegance and fashion are the brand new definition of a mansion

living room 客厅

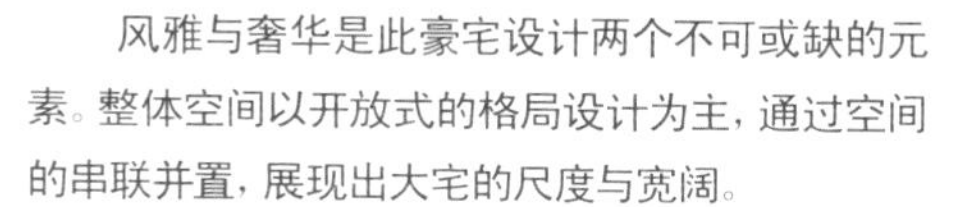

风雅与奢华是此豪宅设计两个不可或缺的元素。整体空间以开放式的格局设计为主，通过空间的串联并置，展现出大宅的尺度与宽阔。

在玄关的设计上，设计师刻意将玄关纳入客、餐厅内，带出了宽阔感。客、餐厅以开放式的设计表现宽敞与气度，由客厅的主墙面至天花板，拉出一道漂亮的弧形曲面，以此整合客、餐厅空间，并化解梁线的突兀。而空间的流畅与现代感，则通过波浪曲线的设计造型呈现，精致时尚的家具及典藏的艺术品相互呼应，反映出优雅的生活模式与品位，并实现住宅空间实践艺术装置的理想。

餐厅区与开放式厨房成一轴线配置，互相呼应，让视觉感与生活功能相兼顾。餐厅挂画主墙面另隐藏一道隔绝内外空间的暗门，增添私人空间的隐秘性。至于私人空间的设计，在主卧前的廊道以简单、干净、舒适的基调，搭配皮革主墙面，突出空间的优雅与大气。整体空间设计既兼具功能性，又彰显奢华的价值，让家秀出时尚气息，却也不失居住的实用性。

living room 客厅

Elegance and luxury are the two necessary elements in the project. The designer adopts the theme of the open layout in the project, presenting the depth and spaciousness by threading and arranging the spaces abreast.

The designer deliberately concludes the entrance into the living room and dining room, in order to obtain the sense of spaciousness that the partition will sacrifice. The designer adopts the idea of open design for the dining room and living room to express the grand manner by extending the main wall of the dining room to the ceiling, and presenting a beautiful arc curved surface, in order to dilute the clumsiness of the beam line. The modern and smooth sense of the space is introduced by the design of curved lines. Delicate modern furniture and artworks echo each other, and reflect the elegant lifestyle and taste, and create the ideal of integrating the Installation art part into the living space.

The dining and the open kitchen echo each other on the line, representing that the designer does take visual enjoyment and life functions. Moreover, the main wall in the dining room hides another secret door, to make the private area even more secluded. As for the design of private area, the hallway in front of master bedroom looks simple, clean, and cozy, along with leather wall, to reflect elegant lifestyle and taste. The whole space also covers life functions to magnify the luxurious value. The home indeed reveals the fashionable style but still maintains the practicability for living.

Picture_TRENDY INTERIOR DESIGN CO; LTD
图片提供_动象国际室内设计

House Data

空间案名 明水大观
设 计 者 谭精忠
参与设计 翟骏、许思宇、何芸妮、黄利画
设计公司 谭精忠室内设计工作室
空间地点 台北市
空间性质 电梯大楼
空间面积 166 m2
空间格局 玄关、客厅、厨房、客房、主卧室、主卧浴室、客浴、客卧
主要建材 染色木皮，刷漆，壁布，木地板，石英砖，夹纱玻璃

Title / Grand Ming-shui
Designer / David Tan
Assistant designer / Jun Zhai, Si-yu Xu, Ynn-ni He, Peter Huang
Design company / TAN CHING-CHUNG STUDIO
Location / Taipei City
Category / Elevator building
Area / 166 m^2
Layout / entrance, living room, kitchen, guest room, master bedroom, master bathroom, guest bathroom, guest bedroom,
Building materials / dyed veneer, paint, wall cloth, wooden floor, quartz tiles, and sheer laminated glass

	02
01	03

01 _ 餐厅与客厅 02 _ 主卧室 03 _ 主卧浴室
01 _ dining room & living room 02 _ master bedroom 03 _ master bathroom

Designer Data

谭精忠
现　任／
动象国际室内设计、大隐设计集团主持人
学　历／
复兴美术工艺科毕业
经　历／

1984	任职于谭国良设计事务所
1989	成立谭精忠室内设计工作室
1999	成立动象国际室内设计有限公司
2002～2003	成立大隐室内设计有限公司上海、台北分公司

简　介／
1984年毕业之后即跨入室内设计界，1990年获意大利TECNHOTEL PACE室内设计奖优胜奖。2006、2007年连续获得亚太区室内设计大奖、日本JCD以及TID台湾室内设计大奖等多个奖项。2007年评为十大样板房设计师之一。

David Tan
Principal of TRENDY INTERIOR DESIGN CO, LTD and TAD ASSOCIATES INC
Professional

1984	Worked at Tan Guo-liang Studio
1989	Established TAN CHING-CHUNG STUDIO
1999	Established TRENDY INTERIOR DESIGN CO, LTD
2002-2003	Establish TAD ASSOCIATES INC., Taipei and TAD ASSOCIATES INC., Shanghai

Education
Fu-Hsin Arts School
Brief Bio
David Tan has entered interior design industry since his graduation in 1984. He gained TECNHOTEL PACE INTERIOR DESIGN 1990 AWARDS. He also gained Asia Pacific Interior Design Awards, JCD Design Awards (Japan) and TID Award (Taiwan) in 2006-2007. Tan received the China Top 10 Sample House/Room Designer Awards in 2007.

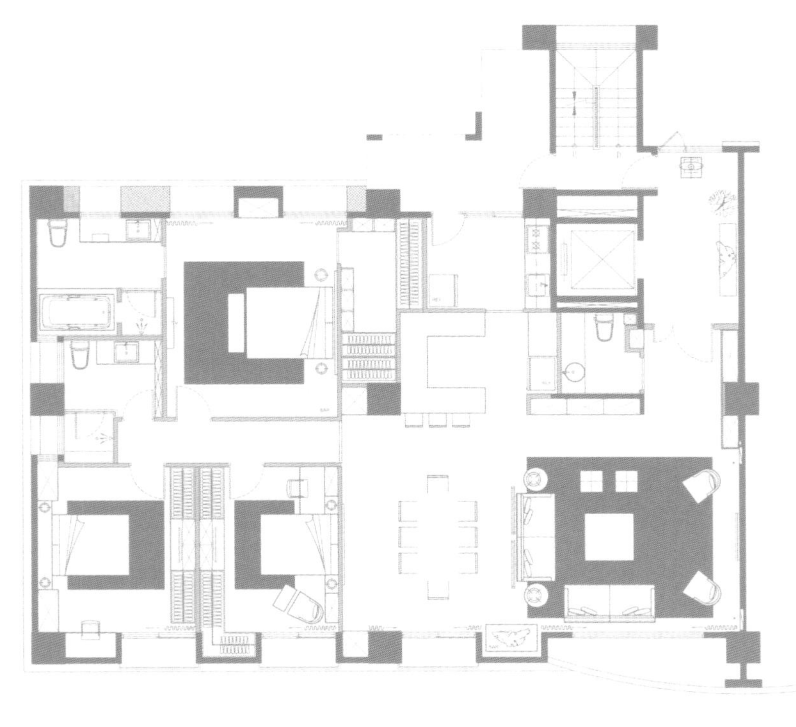

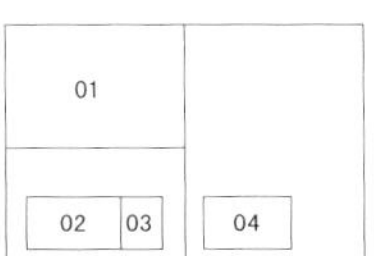

01 _ 主卧室　02 _ 主卧室　03 _ 更衣室　04 _ 主卧浴室
01 _ master bedroom　02 _ master bedroom
03 _ dressing room　04 _ master bathroom

hallway 玄关

由圆成形，有机切割出趣味空间

Formed as a circle, and carefully create interesting spaces

Pictures_Waterfrom Design Photographer_Guomin Lee

图片提供_水相设计 摄影_李国民

进入玄关，就被不完全方正、形似画框的屏风所吸引，从框景中看见前方的弧形，并且让视线引领动线前行。在本案中，格局和动线因应基地条件而发展，设计师以弧形动线具象地串联整个空间的功能，弧形墙面刻画出一道有形的曲线，既区隔空间也连接空间。

意象的表达是设计师在本案中所要强调的重点，由于屋主喜爱阅读，因此在玄关进入客厅的空间，让书柜成为墙面；为融入有机形体的设计概念，以树状的抽象形式，在利落现代的质感中，诉说着自然的温馨。

简化的设计也是本案另一个重点，空间的规划简单不复杂，以生活功能为出发点，通过必要的设计，将功能与形式整合。因此，当空间形式构筑完成时，设计也几乎同时完成。在此空间中，没有多余的缀饰，空间随着结构展开，而充沛的自然光线，就足以表现出空间的丰盈度，以及都会居家所拥有的又一种奢华。

Entering the entrance, a picture-frame like a screen immediately catches visitors' attention. They see through screen, reach the arc and then view further beyond from the screen. In the project, moving lines are developed by the space. The designer strings all the spaces together. The arc wall carves a tangible curve line, both dividing the spaces and connecting the spaces.

The theme of the project is to create a vision. The owner loves reading; therefore the bookshelf serves as wall between the entrance and the living room. The concept of embedding a piece of tangible furniture in the space, looks like a tree, and conveys the nature's image.

Simplicity is another theme of the project. The layout of the space is really simple, only integrating the basic living functions with the design of the rooms. Therefore, once the space constructions are finished, the interior design of each space is also almost done. Therefore, the space does not provide additional decorations. The public area extends along with its structure, presents the abundance of the space and bestows the funky luxury by heartily welcoming the natural light.

living room 客厅

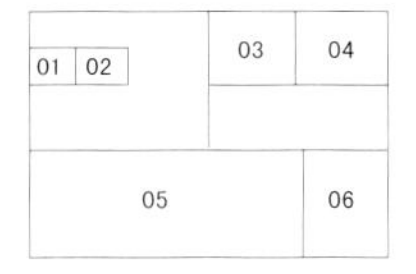

01 _ 玄关与阅读区　02 _ 书墙　03 _ 客厅与餐厅　04 _ 电视墙　05 _ 客厅与阅读区　06 _ 玄关与阅读区
01 _ hallway and reading area　02 _ book wall　03 _ living room and dining room　04 _ TV-wall
05 _ living room and reading area　06 _ hallway and reading area

House Data

空间案名　敦南李宅
设 计 者　李智翔
设计公司　水相设计
共同企划　荷果设计
空间地点　台北市
空间性质　单层住宅
空间面积　165.29 m^2
空间格局　客厅、餐厅、厨房、主卧室、主卧浴室、主卧更衣室、客房、小孩房、客浴、储藏室
主要建材　雾面抛光砖、染白枫木、黑檀木、马赛克、板岩

Title / Lee's house, Dun-nan
Designer / Zhi-xiang Li
Design company / Waterfrom Design
Common business planning / hercore
Location / Taipei City
Category / Ground floor apartment
Area / 165.29 m^2
Layout / Living room, dining room, kitchen, master bedroom, master bathroom, master dressing room, guest room, guest bathroom, nursery, and storage room
Building material / matte polished tiles, stained white maple, and ebony, slate

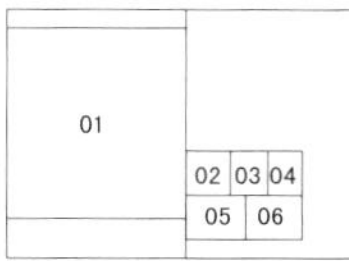

01 _ 客厅与阅读区　02 _ 阅读区　03 _ 客厅　04 _ 浴室与阅读区
05 _ 主卧室　06 _ 阅读区
01 _ living room and reading area　02 _ reading area　03 _ living room
04 _ bathroom and reading area　05 _ master bedroom　06 _ reading area

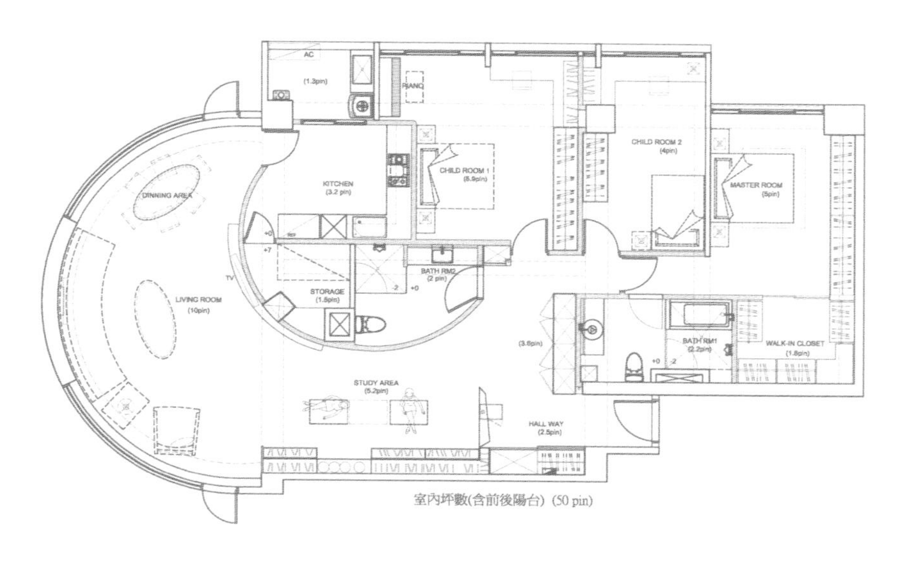

室內坪數(含前後陽台) (50 pin)

Designer Data

李智翔

现　任／

水相设计　设计总监

学　历／

1998 ~ 1999　丹麦哥本哈根大学研修毕业

1997 ~ 2000　美国纽约布拉特学院（Pratt Institute）室内设计硕士

经　历／

2008　水相设计 设计总监

2004 ~ 2008　荷果设计 设计总监

2001 ~ 2003　MMoser 香港商穆氏设计集团

2000 ~ 2001　李玮珉建筑师事务所

简　介／

重新再定义空间，创造真正量身订制的个性住宅，富含幽默感的设计语汇，是李智翔设计师不断尝试在空间注入的特质。近期作品更追求“自然光与线条比例”的相对关系，以更内敛但不含蓄的手法处理每一项细节，创造出独特的设计风格。

Zhi-xiang Li

Waterfrom Design, Design director

Education

1998–1999　The University of Copenhagen, Research programmes

1997–2000　Pratt Institute, New York, USA. Mater of Interior Design

Professional

2008–present　Waterfrom Design, Design director,

2004–2008　Hercore Design, Design director,

2001–2003　MMoser

2000–2001　LWMA

Brief bio

Li Zhi-xiang attempts to instill the space with the refreshing quality by redefining the space, creating the tailor-made personal houses, and employing humorous vocabulary. His recent works pursue the relative relationship between "natural light and linear proportion." And Li handles every detail in a restrained but not reserved way, in order to create absolutely unique designs.

形随功能而设计的生活容器

A residential container designed for flexible functions

living room 客厅

resting room 休闲室

设计师重视空间线条及比例，认为空间线条越简洁越能达到放大空间的效果。但不能为了空间线条而牺牲空间功能，因此设计师特别发展出隐藏式收纳设计，将柜体隐身在空间的立面，加上无手把的设计，让柜子完全看不出所在。柜面的材质及造型变化，让收纳柜不只用来收藏物品，同时也提升了空间的质感，并营造了空间的风格。

空间依屋主需求规划了3间卧室，设计师将长形客厅切割，规划开放式和室，并连接开放式餐厅，原本无用的走道在移动了卧房的隔断墙后，有了不同的面貌。玻璃展示柜，让走道也成为重要的展示、收纳空间。除此之外，灯光的变化运用，也让空间展现出不同的情调。

设计师特别选择浅色作为空间主色，以简洁的线条为架构，通过灯光及材质的变化，让空间展现出质感及品位；设计师除了以其擅长的隐藏式收纳创造出强大的收纳功能，还结合自动化控制系统，整合音响、电视、灯光及窗帘，让生活更为便利。

In the emphasis of spatial lines and scale, the designers felt that a better effect of expanded spaciousness can be achieved by cleaner spatial lines. But one cannot sacrifice functionality for the sake of cleaning up the lines. Hence, a special design philosophy of hidden storage spaces is adopted where storage units were incorporated into spatial surfaces. Cabinets can be designed without handles to hide them from view, while its surface and external shape can be altered such that the cabinet can not only function as a storage unit, but can also be used to improve the spatial quality and create a unique theme for the compartment it exists in.

Three bedrooms were planned according to the needs of the residents. The designer dissected the rectangular living room to form an open washitsu that connects the open-style dining room. The scarcely used walkway gains a new appearance after the designer shifts the dissecting wall between the bedrooms. With its new glass display closet, the walkway also becomes an important compartment for the storage and display of items. Utilization of light dynamics also gives the compartments different tones.

Light colors are chosen to create the unit's theme. By using clean lines as the framework and utilizing the dynamism between lighting and surface materials, the modified compartments and rooms are able to express novel textures and tastes. Besides the signature use of hidden storage units to create a large storage volume, the designer also integrated an automatic control scheme for music players, television, lighting and curtains, giving life a greater ease.

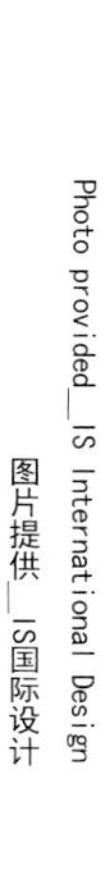

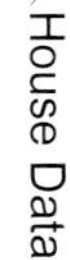

设 计 者　陈嘉鸿
设计公司　IS国际设计
空间地点　新北市
空间性质　大楼
空间面积　165 m²
空间格局　玄关、客厅、餐厅、和室、厨房、客用浴室、工作阳台、主卧室、更衣室、主卧浴室、小孩房、长亲房
主要建材　白栓木木皮、高级壁布、意大利进口板岩瓷砖、浮雕橡木、染灰海岛型木地板、雪白银狐大理石、白洞石大理石、黑金峰大理石、南非黑大理石

Designer / Chia-Hung Chen
Company / IS International Design
Location / Taipei
Type of space / Large flat
Area / 165 m²
Layout / Front porch, living room, dining room, washitsu, kitchen, guest bathroom, working balcony, master bedroom, dresser, main bathroom, children's room, elder's room
Main materials / White cherry bark, quality wallpapers, high grade Italian slate tiles, high grade embossed oak wood island style gray dyed flooring, high grade snow fox marble, high grade white cave marble, high grade black gold-peaks marble, high grade South African black marble

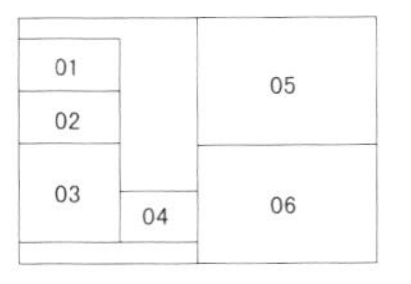

01 _ 餐厅　02 _ 餐厅　03 _ 走道　04 _ 客厅　05 _ 客厅　06 _ 休闲室
01 _ dining room　02 _ dining room　03 _ hallway
04 _ living room　05 _ living room　06 _ resting room

Designer Data

陈嘉鸿

现　任／

IS国际设计 主持设计师

简　介／

凭借天生的空间感和独特的美学品味，投入室内设计的领域，擅于运用垂直及水平线条来捕捉空间向度，并常以隐藏式的收纳方法及家具的搭配，打造出独特的居家风格。著有《室内设计师陈嘉鸿的隐藏学》。

Designer Data

Chia-Hung Chen

Current employment

Lead Designer, IS International Design

Resume

Chen combines an innate acute spatial sense with a unique set of aesthetics, and is fully devoted to the field of interior design. His precise handling of scale and skilled use of vertical and horizontal lines allows an accurate depiction of spatial direction. The hidden style of storage methods combined with the furniture creates a unique residential style.

Publications

Hiding Methods of Interior Designer Chia-Hung Chen, published by MyHomeLife

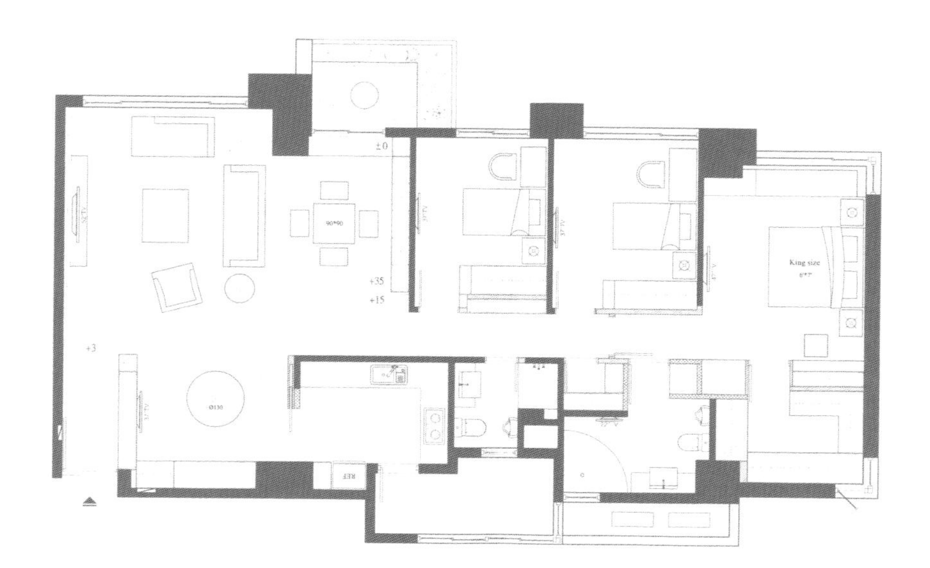

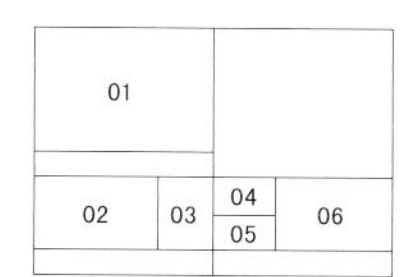

01 _ 主卧　02 _ 男孩房　03 _ 浴室　04 _ 主卧
05 _ 浴室　06 _ 女儿房
01 _ master bedroom　02 _ gent's room　03 _ bathroom
04 _ master bedroom　05 _ bathroom
06 _ lady's dressing chambers

有机空间，流畅而宽广的尺度

Organic space, with smoothness and spaciousness

dining room and balcony 客厅与阳台

本案为旧屋改建，屋主委托设计师为其父母的老家重新整修。由于是长辈的居住空间，在空间质量与细节的设计上，坚持弹性便利与无障碍原则；在平面配置上，放弃垂直水平思考，而以有机形体、居住者意识中心的概念，解决对空间尺度的安排。

设计师在公共空间与私人空间的交界处，运用斜面角度，让有限的空间看起来更大更舒适。有机的概念同样也应在玄关空间，以铁件框架设计衣帽间及鞋柜，表面使用透气面网材质包覆，同时通过斜角所构成有机空间，使空间发挥最大的可变性效能。另外，空间整洁便利、通风流畅，以及行动方便、减少高低差、地板几乎无障碍等，也都包含在设计考虑之中。

在主卧室的空间设计上，以透明玻璃作为睡眠区和卫浴空间的隔断，穿透性的设计手法，让空间感更通透。在整个空间的设计上，整合出视觉的流畅性，让人久看不腻。

The project is a renovation of an old house. The client asked the designer to renovate his parents' house. Therefore, the designer focuses on convenience and accessibility, and centers upon the senior dwellers' need.

The designer employs the slope to connect public and private areas, so that the limited space will look more airy and more comfortable. The same organic concept of slope is adopted in the entrance. The cloakroom and the shoe cabinet are made of iron net, and wrapped with venting material. Moreover, the organic place constituted by the slope angles maximizes the flexibility of the space. Moreover, the easiness to clean, ventilation, narrowing the heights and the accessibility floor for safety are all taken into account in the design.

In the master bedroom, the wall segmenting the sleeping area and bathroom is made of the light glass, so the whole room looks more spacious. Overall, the design of the whole space brings about visual smoothness and therefore makes the space always look refreshing.

图片提供_陆希杰设计事业有限公司 摄影_Marc Gerritsen

Pictures_ C.J Studio Photographer_ Marc Gerritsen

living room and balcony 客厅与阳台

House Data

空间案名　板桥蔡宅
设 计 者　陆希杰
参与设计　曹车政、刘致佑、郑家皓
设计公司　陆希杰设计事业有限公司
空间地点　台北板桥
空间性质　公寓
空间面积　154 m^2
空间格局　客厅、厨房、餐厅、书房、客房、客浴、主卧室、更衣室、主卧浴室
主要建材　栓木地板、黑花岗烧面石材、烤漆玻璃、条纹玻璃

Title / Cai' s house, Banqiao
Designer / Shi-chieh Lu
Assistant designer / Che-zheng Cao, Zhi-you Liu, Jia-hao Zheng
Design company / CJ STUDIO
Location / Banqiao, Taipei County
Category / Condo
Area / 154 m^2
Layout / Living room, kitchen, dining room, study, guest room, guest bathroom, master bedroom, master bathroom, and dressing room
Building material / solid wood flooring, burning surface of the black granite stone, stoving varnish glass, and reeded glass

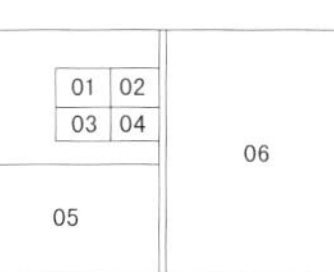

01 _ 客厅　02 _ 餐厅　03 _ 走道　04 _ 餐厅与走道　05 _ 餐厅与玄关　06 _ 走道看向书房
01 _ living room　02 _ dining room　03 _ aisle　04 _ dining room and aisle
05 _ dining room and hallway　06 _ study towards the living room

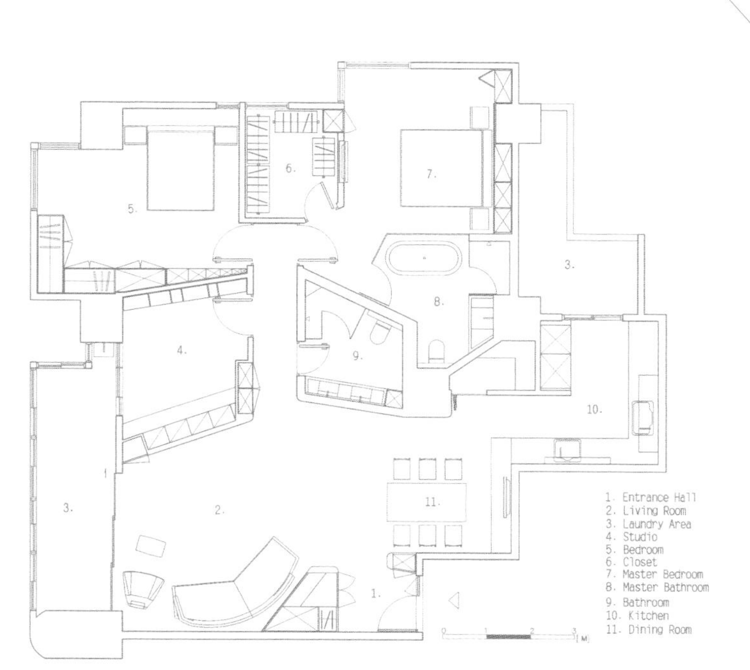

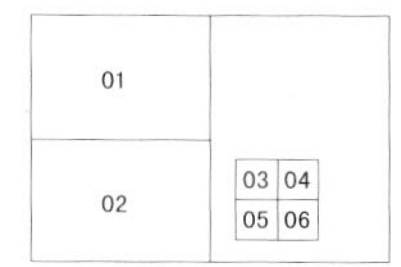

01 _ 主卧与主卧浴室　02 _ 主卧与主卧浴室　03 _ 客房　04 _ 主卧浴室　05 _ 厨房　06 _ 浴室
01 _ master bedroom and master bathroom　02 _ master bedroom and master bathroom　03 _ guest room
04 _ master bathroom　05 _ kitchen　06 _ bathroom

Designer Data

陆希杰
现　任／
CJ STUDIO 主持人
学　历／
东海大学建筑系毕业。英国AA建筑联盟硕士
简　介／
从事建筑及室内设计、家具设计、产品设计以及相关领域的研发，现兼任交通大学建筑研究所助理教授，著有《锻造视差》。代表作品有："窦腾璜及张李玉菁wum 概念店"；"窦腾璜及张李玉菁台中Tiger city概念店"（2005年获得日本JCD设计大奖）；"Aesop微风店"（2006年获得日本JCD设计大奖、国际室内设计联盟IFI 2007设计金奖）；"敦南国美蔡宅作品"（获得台湾室内设计大奖TID Award 2007金奖）。入选美国室内设计杂志《INTERIOR DESIGN七十五周年特集》中五位具潜力设计师之一。2008年，参展"第十一届威尼斯建筑双年展"和台北当代艺术馆"第六届城市行动艺术节"。

Shi-chieh Lu
Professional
CJ STUDIO, design director
Education
Master of Architectural Association School of Architecture, UK
Department of Architecture, Tunghai University
Brief Bio
Shichieh Lu is an adjunct assistant professor in Graduate Institute of Architecture, NTCU, specializing in interior design, furniture design and product design. He authors
Forging parallax, published in 2003. Lu's masterpieces include
•Stephanie Dou & Changlee Yugin's concept store, WUM
•Stephan Dou & Changlee Yugin's concept store in Taichung, Tiger City (won 2005 JCD Design Award)
•Aesop in Breeze Store (won 2006 JCD design award and IFI 2007 golden prize)
•Cai's house in Dun-nan Guo-mei (won TID Award 2007 golden prize) .
Lu, listed as one of the five promising architects in Interior Design Magazine's special issue for its 75 anniversary, participated in the 2008 Biennale Architecture 11th International and 2008 MOCA TAIPEI-dark city

living room and dining room 客厅与餐厅

开放风景，坐拥一室绿意

Open your window and embrace a scene of greenery

结合屋主姓氏英文的首字母，以及整个空间建筑结构形式，"L" 成了贯穿本案设计的概念。考虑到房子的采光良好，在尽可能保留四周景观的前提下，成就了开放的公共区域，而整个空间更由3个不同大小的L形区块组合而成。

最主要的空间设定在由客厅、餐厅以及书房组成的 "L" 形区域，大面积的开窗将室外的竹林景致引入室内，明亮通透以及被绿意围绕的感觉营造出舒适的氛围。通过打破空间界线的设计手法，以及浅色系家具的使用，衬托出户外的绿景，让空间感更为放大。

主卧室、起居室分别以活动式拉门与客厅区隔。同样呈现 "L" 形的卧房在拉门开启时，斜角窗户的景观可与客厅窗外的风景相呼应，并且创造出更为宽阔的视野。而房内的畸零空间，则通过收纳柜的配置强调出功能性，洗手台配置在床头后方，既简单区隔空间，也让视线不受阻，换来更为自由的视野。通过融合室内与户外风景，打造出清新素雅的人文居家。

By incorporating the initial of the surname of the house-owner onto that of the whole design structure, the letter "L" has become the central concept of this project. Considering the fact that natural lighting comes in naturally to the apartment, an open public area is created under the premise of trying to retain as much scenic view as possible. As for the area itself, it is made up of three L-shaped blocks of different sizes.

The main area comprises of the L-shaped zone encompassing the living room, dining room and study room. Large window allows the inhabitants to enjoy the greenery of the bamboo forests outside. A comfortable and layback atmosphere is generated by the feeling of being surrounded by a vast forest in a pleasant day. Along with the use of light-colored furniture to complement the foliage outside, the design breaks down boundaries in spaces, to create an enlarged and airy sense of space.

The master bedroom and living room are divided by a sliding door. When the sliding door is pulled, the adjacent window on the bedroom, which is also L-shaped, displays a beautiful landscape in par with the views from the living room window. This helps to produce an even more spacious feel. As for the fractional spaces within the apartment, storage cabinets are placed in these areas to increase the functionality of the space. The sink is located at the back of the bedside cupboard. The end result is that spaces are divided without difficulty; views are not blocked; and a more expansive view is resulted. By fusing interior designs with outside scenery, the designer has created a tranquil and elegant home.

Pictures_Materiality Design Photographer_ Kyle Yu
图片提供_石坊空间设计研究事务所 摄影_Kyle Yu

living room 客厅

House Data

空间案名 拢翠
设 计 者 郭宗翰
参与设计 陈婷亮
设计公司 石坊空间设计研究
空间地点 台北
空间性质 电梯大楼
空间面积 151 m²
空间格局 客厅、餐厅、厨房、书房、卧室、浴室
主要建材 木地板、铁件、染色木皮、盘多磨、石材、玻璃、砖

Case Name / Long Tsui
Designer / Tsung-Han Kuo
Co-designer / Ting-Liang Chen
Company / Materiality Design
Location / Taipei City
Nature of the apartment / Condominium
Area / 151m²
Outline / living room, dining room, kitchen, study room, bedroom, bathroom
Main building material / wood floors, iron, wood staining, Pandomo, stone, glass, brick

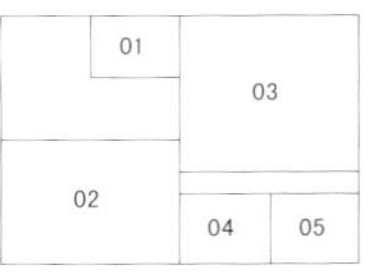

01 _ 客厅 02 _ 客厅 03 _ 书房与餐厅 04 _ 书房 05 _ 餐厅与客厅
01 _ living room 02 _ living room 03 _ study and dining room
04 _ study room 05 _ dining room and study room

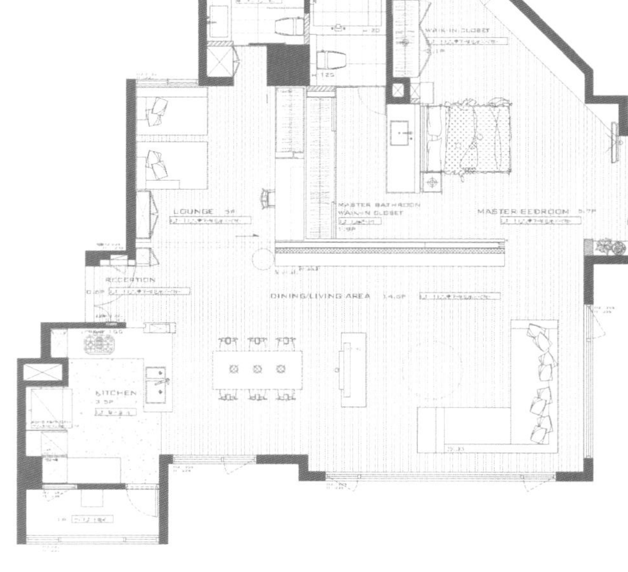

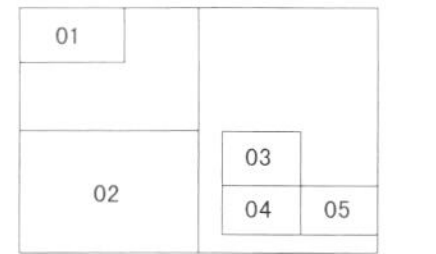

01＿客厅　02＿餐厅与书房　03＿卧房　04＿餐厅　05＿卧房
01＿ living room　02＿dining room and study room　03＿bedroom
04＿dining room　05＿bedroom

Designer Data

郭宗翰
现　任 /
石坊空间设计研究　设计总监
学　历 /
1997 ~ 1999　英国伦敦艺术大学空间设计系学士
1999 ~ 2000　英国北伦敦大学建筑设计系硕士
经　历 /
2000 ~ 2002　香港商穆氏设计　设计师
2002　石坊空间设计研究　设计总监
2005　实践大学设计学院兼职讲师

Tsung-Han Kuo
Current Position
Director, Materiality Design
Education
1997-1999　University of the Arts London, BA in Spatial Design
1999-2000　University of North London, UK , MA in architectural design
Experiences
2000-2002　Designer , Moody's Commercial Design, Hong Kong
2002　Director, Materiality Design
2005　Lecturer at College of Design, Shih Chien University

轴线串联，打造大尺度居家

“Flexible routes”: creating a spacious home

living room 客厅

living room and dining room 客厅与餐厅

空间的宽敞是来自于轴线的串联。本案位于台北市高级地段的住宅区，在方正的格局中，设计师通过对轴线的掌控，让每一个空间可以通过线性的配置手法加以串联，而空间与空间则以书房为中心，架构出既独立而又具层次感的居住环境。

在客厅与餐厅的区块，设计师以中岛作为区隔，中岛半遮蔽的功能维持了客厅与餐厅的独立性，却又能以视线穿透，创造出宽敞的空间感。书房安置在公共区域与私人区域的中间，经由此处，让屋主可视需求调整空间，让空间使用方式更为灵活。

方正的格局是本案的特色。虽然空间分隔为几个区块，但设计师善用轴线串起每一个空间，因而将空间的尺度放大。功能的设定都被隐藏在设计细节之中，因而打造出整洁的立面。餐厅与书房均搭配了不同个性的吊灯，既打破空间的呆板，也为空间演奏出不同的旋律，丰富了空间的个性与表情。

The spacious room comes from the interconnection of the routes. The property is located at a classy, low–density residential area in Taipei. Within the square layout, the designer takes control of every route, so that every space is connected through 'linear configuration". This in turn creates a living space that is both independent in nature as well as having a sense of hierarchy.

As for the block which contains the living room and the dining room, the designer used "mid–island" as the divider of the two dimensions. The semi–masking function of the divider helps to maintain the independency of the living room and dining room from each other, but in the same time creating a spacious space in which the inhabitants can move freely. This is done by the use of "visual penetration technique". The study room is located in between the public and private zone; as one passes here, he/she can decide which direction to go depending on his/her needs. The versatility of design is fully demonstrated here.

The use of square layout is the characteristic of this case. Even though the living space is divided into many smaller sections, each space is interconnected through the use of flexible routes. This creates the illusion of a larger space. All the functional components are hidden within the design. This intricate designing technique in turn creates a clean outlook. Different chandeliers are used in the dining and study rooms. This breaks down the rigidity of the space and makes the design full of life and expression.

Photo courtesy_MPI design
图片提供_MPI design

House Data

空间案名　一庄
设 计 者　王玉麟
参与设计　陈亮瑄
设计公司　MPI设计公司
空间地点　台北市 天母
空间性质　电梯大楼
空间面积　150 m²
空间格局　玄关、客厅、餐厅、厨房、书房、主卧室、卧房×2、主卧更衣室、主卧浴室、浴室×3
主要建材　钢件、石材、玻璃、木皮、铁件、硅酸钙板

Case Name / Yi-Zhuang
Designer / Yu-Lin Wang
Co-designer / Liang-Hsuan Chen
Company / MPI design
Location / Tianmu, Taipei City
Nature of the apartment / Condominium
Area / 150 m²
Outline / entrance, living room, dining room, kitchen, study room, master bedroom, bedroom x2, master dressing room, master bathroom, bathroom x3
Main building material / Steel, stone, glass, wood, iron parts, calcium silicate board

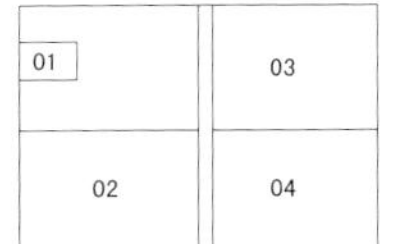

01 _ 书房与餐厅　02 _ 主卧室　03 _ 书房　04 _ 书房
01 _ study and dining room　02 _ master bedroom
03 _ study room　04 _ study room

01		
02	03	04 / 05

01 _ 餐厅与书房　02 _ 卧房　03 _ 卧房
04 _ 主卧室　05 _ 浴室
01 _ dining room and study room　02 _ bedroom　03 _ bedroom
04 _ master bedroom　05 _ bathroom

Designer Data

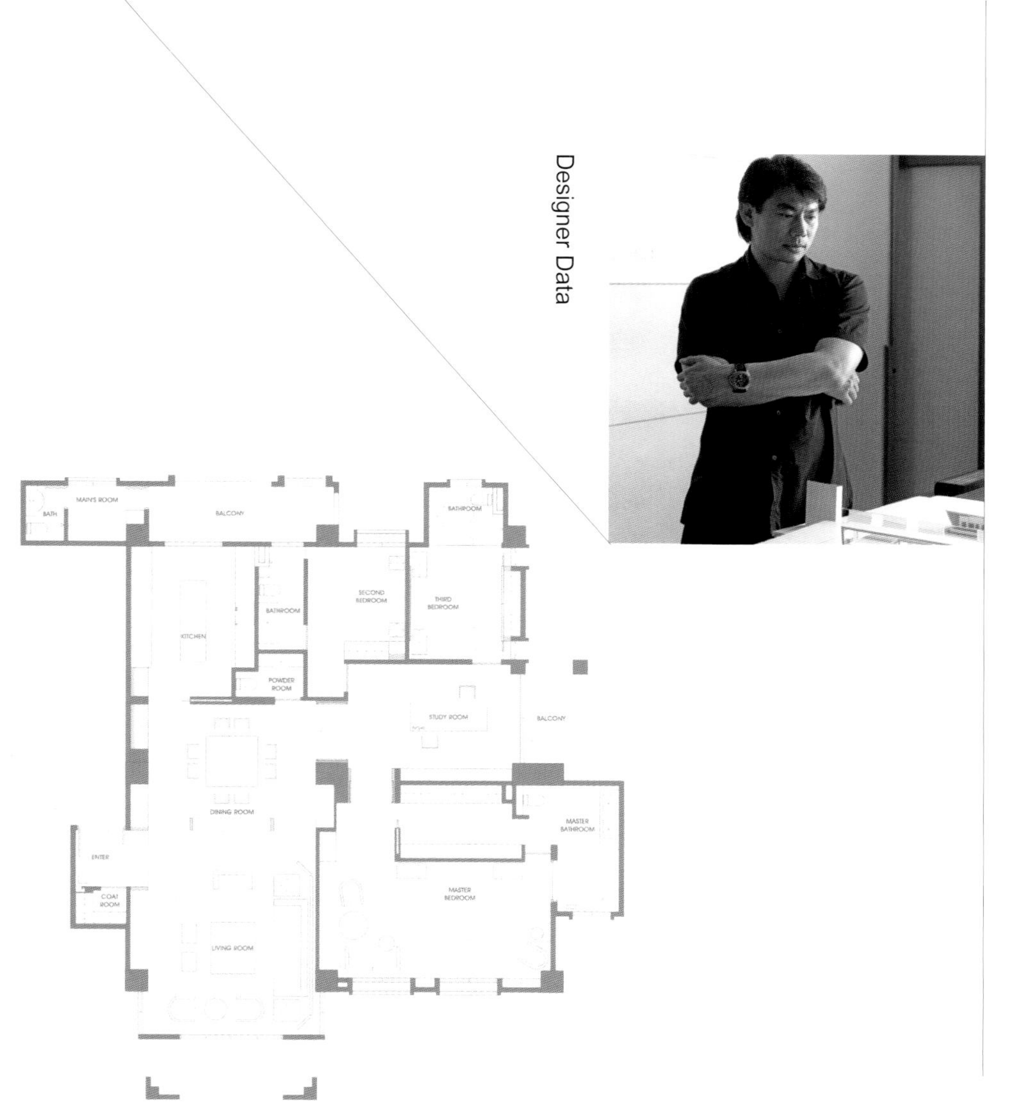

王玉麟

现　任／

MPI设计公司执行长

学　历／

法国巴黎艺术学院毕业

经　历／

MPI设计集团(美国纽约)IFI国际设计师联盟会员

简　介／

2003～2005年连续获得三届日本JCD设计大奖，2007、2008年获选为TID台湾设计奖TID奖、公共空间类金奖以及临时建筑类TID奖。现为台湾观光旅馆质量监评委员、中原大学室内设计系助理教授

Yu-Lin Wang

Current Position

Executive, MPI-DESIGN

Education

Ecole des Beaux-Arts, France

Experiences

MPI-DESIGN GROUP NEW YORK USA

IFI international designer's union, affiliate

Brief

JCD Design Award (in Japan) for three consecutive years, from 2003-2005; TID Design Award (Taiwan), Gold Award for Public Space Design, TID Award: Provisional Building Award, for 2007 and 2008; currently takes the position as the Taiwan Hotel Quality Assessment Committee Member as well as being the Assistant Professor of Chung Yuan Christian University.

dining room and living room 餐厅与客厅

引介自然，解放空间
Inducing a naturally free space

family room 休憩室

living room 客厅

本案基地位于坐拥整片山林的绝佳位置，由于屋主单身，在空间的规划与功能设计上可以更为单纯。为了替居住者创造舒适、放松的生活环境，设计师将室内空间与户外的绿意视为一个整体，再通过空间框架的拆解，展现出无拘无束的自由度。

讲究与自然的呼应关系，这种关联性的重视也落实于空间的规划上。电视墙左右各设计两道拉门，可分别进入主卧室以及书房，空间既可完全开放，也能以拉门区隔，保护了私人空间的隐私。在书房的设计上，可看出设计师力图创造“留白”的用心。卧榻佐以窗外的绿意，伴以纯白的墙面，在此处停留，仿佛可随时对自己忙碌的生活喊声“暂停”。

借助自然景色的引介，以及空间格局的规划，将空间尺度释放至最大。框架的拆解，让生活空间更具悠游自在的特质，营造出真正安心休闲的居家环境。

This project is situated on an excellent location overseeing a forested hillside. Since the property owner is single, spatial and functional planning can be made simpler, providing the resident with an unfettered living environment. Designer Chung-Han Tang takes the view that both the interior space and the external green landscape exist as a single entity. His dissection of the spatial framework and compartments results in the purest expression of freedom.

The emphasis on a connected relationship with the natural landscape is manifested in the spatial planning. The two sliding doors on the two sides of the television panel wall allow entry to the master bedroom and the study. The rooms can either be opened to all or closed to retain the user's privacy. A deliberate and creative "white minimalism" of the designer is evident in the study. The futon couch is positioned beside the green scenery outside the window whilst facing a clean white surface. A short rest here to break away from the busy schedule of modern life can be made here at any moment.

Hence, by inducing natural scenery and planning of spatial layouts, this project makes the most out of the spatial limitations. Through dissecting the spatial framework and compartments, a character of freedom is given to the living quarters, creating a residence where one can truly rest with a peace of mind.

Pictures_Design Apartment Creation
图片提供_近境制作

House Data

空间案名 绿光宅
设 计 者 唐忠汉
设计公司 近境制作
空间地点 台北
空间性质 电梯大楼
空间面积 150 m²
空间格局 客厅、餐厅、厨房、主卧室、主卧卫浴、书房、客浴
主要建材 石片、柚木、杉木、黑铁件

Project Title / Greenlight Residence Type L.
Designer / Chung-Han Tang
Company / Design Apartment
Locale / Taipei City
Type of space / Multi-storied building with elevators
Area / 150 m²
Layout / Living room, dining room, kitchen, master bedroom, master bathroom, study, guest bathroom
Main materials / Stone facing, teak, Chinese fir, steel

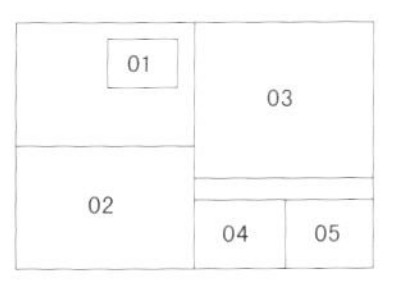

01 _ 餐厅 02 _ 书房 03 _ 餐厅与厨房 04 _ 休憩室与书房 05 _ 主卧
01 _ dining room 02 _ study room 03 _ dining room and kitchen
04 _ family room and study 05 _ master bedroom

Designer Data

唐忠汉
现　任／
近境制作设计总监
学　历／
中原大学室内设计学系
简　介／
2009年获台湾室内设计大奖居住空间、商业空间类 TID奖

Tang Zhonghan
Current position
Chief Designer
Professional training
Interior Design, Chung Yuan Christian University
List of Honors
2009 Taiwan Interior Design Award–Residential Space Category
2009 Taiwan Interior Design Award–Commercial Space Category

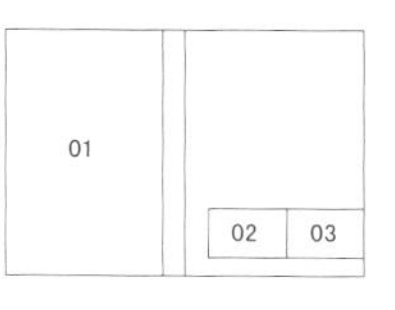

01 _ 客厅　02 _ 主卧　03 _ 主卧浴室
01 _ living room　02 _ master bedroom　03 _ master bathroom

都会时尚与自然休闲共构舒适居家

Building a home where urban fashion and nature come together

entrance 玄关

秉持着一贯的"不设限"设计理念，设计师打造出都会时尚调性与自然休闲情趣共融的居家风格，于朝气蓬勃中展现出一种特别的宁静祥和。在台北内湖区，如此繁华而忙碌的都市里，设计师用心为屋主安排了一处最舒适的避风港。

本案的特色在于使用了相当多的自然素材，如玄关处以洗石子铺陈廊道，搭配灯光精心营造的玄关端景，连玄关柜体的把手都以自然造型的木材设计而成，一入门就让人感受到富有禅意的舒适气氛。而木材质感的表现是本案重点之一，从墙面、柜体、地板，几近一体成形的木作，在投射灯光下，映照出其纹路的优雅，温和的色调更让空间显得宽敞自然。

然而在自然之中，前卫的设计则为空间带来了活力。卫浴空间利用黑色大理石，营造出奢华时尚的都会风情；而客厅空间的电视墙上方打上蓝色发光二极管灯光，带来前卫科技质感；再搭配普普风的家具与灯饰，整体空间的风格就活泼了起来。

The main philosophy of this design is "no limit". In this case, urban fashion and nature come together, creating a sense of tranquility within the vigor of the youth. The property is located at Neihu, Taipei. At this bustling and busy place, the designer has created a most comfortable haven.

The specialty of this case is that it uses a lot of natural materials. For example, washed stones are used at the entrance to pave the aisle. This matches look well with the elaborate lighting at the entrance. Furthermore, even the handle of the cabinet at the entrance uses natural-shaped wood as its main material, so that, when one walks into the door, the person can immediately be surrounded by a "Zen"-filled atmosphere.

You can feel the wood textures all the way from the wall, the cabinet, to the floor. This complements well with the use of lighting, which emphasize the natural and elegant patterns on the wood's surface. In addition, the use of warm colors also makes the whole place look more natural and spacious.

However, within these natural surroundings, the avant-garde design has also added vitality to the space. The bathroom uses black marble to create a sense of urbane extravagance. Furthermore, blue LED light has been used to light the top of the TV wall at the living room. This creates something of technical sleekness. This complements look well with the pop-styled furniture and lighting, which all helps to liven up the space.

main wall 主墙

图片提供_艺念集私
Pictures_D-A design Ltd.

House Data

空间案名 内湖 住宅案
设 计 者 黄千祝、张绍华
设计公司 艺念集私空间设计有限公司
空间地点 台北市
空间性质 电梯大楼
空间面积 148 m^2
空间格局 玄关、客厅、餐厅、厨房、浴室、主卧房、更衣间、主卧浴室
主要建材 砂岩石、卵石、洗石子、清水模、黑铁件、进口壁纸、大理石、立体皮板、木地板、玻璃、不锈钢材、特殊漆

Case Name / Neihuapartment case
Designer / Anita Hwang&David Chung
Company / D-A design Ltd.
Location / Taipei City
Nature of the apartment/Condominium
Area / 148m^2
Outline / entrance, living room, dining room, kitchen, bathroom, master bedroom, dressing room, master bathroom
Main building material / sand rock, cobble stone, wash stone,fair faced concrete, gun metal, imported wallpaper, marble, stereoscopic leathered plate, wood floors, glass, stainless steel, special paint.

01	03
02	04

01 _ 餐厅 02 _ 客厅 03 _ 玄关端景 04 _ 沙发区
01 _ dining room 02 _ living room
03 _ entrance side view 04 _ sofa area

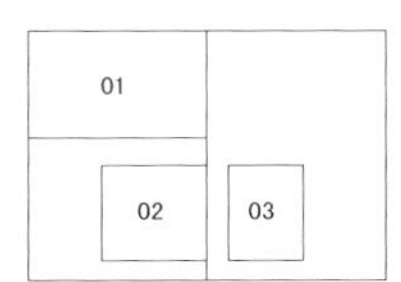

01 _ 主卧　02 _ 主卧卫浴　03 _ 客用卫浴
01 _ master bedroom　02 _ master bathroom
03 _ guest's bathroom

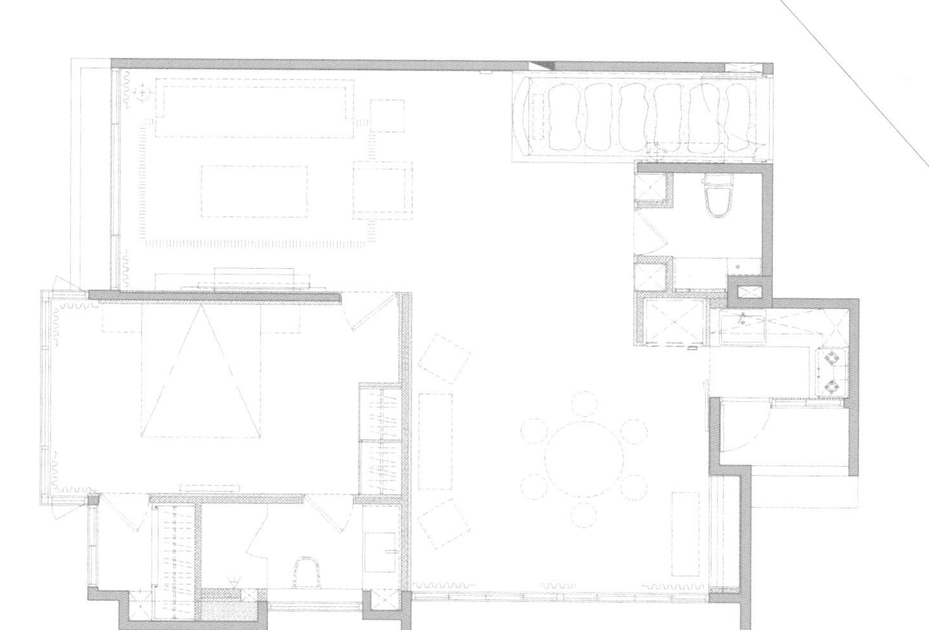

Designer Data

黄千祝
现　任／
艺念集私空间设计有限公司 创意总监
学　历／
协和工商职业学校室内设计科毕业，美国普莱斯顿大学高级工商管理硕士

张绍华
现　任／
艺念集私空间设计有限公司　执行总监
学　历／
复兴美工空间设计专业毕业

简　介／
从不设限自我风格，善用素材本身的美感，利用不同材质及配色展现空间变化。擅长使用建材的特性，特别专注于铸铁材质的运用，将其自然呈现于空间设计中。创意的混搭装置美学，营造出具有个人特殊风格及时尚的空间，让居住者得以释放压力，又呈现出空间的无限可能性。

Anita Hwang
Current Position
D–A design Ltd. Chief Creative Officer
Education
Unit of Interior Design, Xie–He High School of Industry and Commerce, Preston University / EMBA

David Chung
Current Position
D–A design Ltd. Chief Executive Officer
Education
Graduated in Division of Space Design, Unit of Art and Craft, Fu–Hsin Trade and Art School

Brief
Never limit oneself to a certain style, we are able to show off the natural beauty of the materials themselves. By using different materials as well as color complementation, the variation in the spatial design is fully exhibited. We are also good at taking advantage of the special properties of the materials themselves. A special emphasis is placed on the use of casted irons, and naturally presents them in our designs. Finally, a fashionable space is created where unique and personalized styles are valued by mixing creativity and aesthetics. This allows the inhabitants to release the tension in their lives as well as creating designs with infinite possibilities.

奢华内涵，见于设计细节

Luxuries are found in the details

Pictures_Da-Yuan Interior Design Ltd.
图片提供_大院室内装修有限公司

功能性与视觉美感的整合，对于设计师始终是高难度的挑战。在本案中，设计师以居住的本质为出发点，思索空间的布局安排以及功能性配置。通过两条垂直中轴线的定位，让家居空间的格局更贴近居住者的生活习惯，从而营造出简约而舒适的环境。

公共空间由十字轴线所定位。一条轴线从厨房延伸至书房空间，与之垂直的另一条轴线则连贯整个公共空间。轴线不但成功地串联起空间，更让格局有了更精准的定位。餐厅与厨房采用开放式设计，因视野可延伸至书房而创造出更为宽广的空间感。连贯客厅、书房与餐厅的轴线，既串联了公共空间，使之成为一个整体，同时又成为连通各空间的过道。

设计师的细腻还可见于私人空间的规划。主卧室空间有完整的配备，空间利落的切割将其功能做了更清楚的区分。通过清晰的规划，功能被安置在每一个空间中，让居住的品质被提升至顶级，居住者因而能体验并享受到量身订制的奢华内涵。

The integration of functionality and visual aesthetics is always the most difficult task for designers. In this case, the essence of living is adopted as the starting point by the designer, where he thinks carefully about the arrangements of the layout as well as the alignment of functional elements. Through the positioning of two vertical design axis, the spatial layout of the property become closer to the lifestyle of the inhabitants, thereby creating a simple yet comfortable environment.

The public space is positioned by a cross-shaped axis. The positioning of the axis can be found in the linearity of space which extends from the kitchen all the way to the study room, as well as from another axis which is perpendicular to the aforementioned axis, which links the entire public space. The use of design axis not only successfully links each space, but also allows the layout to be configured properly. A sense of a broader space is created by the employment of open design in the dining room and kitchen, where the visual distance is extended all the way to the study room. The design axis which links the living room, the study room and the dining room helps to make the public space a single entity, as well as acting as a passageway which connects each space.

The designer' s attention to details can also be found in his planning for the private space of the apartment; although the master bedroom is rich in functional designs, but the clear-cut design allows the functionality part to be partitioned more clearly. Through careful planning, functionality is assigned to their individual space, allowing the living standard to be upgraded to the highest level, where the inhabitants can fully enjoy themselves amongst the extravagance of the design.

House Data

空间案名　大直明水御
设 计 者　赵玺、柯元杰
参与设计　徐尹之、张雅雯、陈奕轩
设计公司　大院室内装修有限公司
空间地点　台北市大直
空间性质　电梯大楼
空间面积　145 m²
空间格局　客厅、餐厅、厨房、书房、主卧室、主卧浴室、主卧更衣室、浴室
主要建材　橡木、镀钛钢板、抛光石英砖、皮革纹美耐板、壁纸、茶色镜

Case Name / Dajhih Mingshuiyu
Designer / Hsi Chao, Yuen-Chieh Ke
Co-designer / Yin-Chih Hsu, Ya-Wen Chang, Yi-Hsuan Chen
Company / Da-Yuan Interior Design Ltd.
Location / Da-Tze District, Taipei City
Nature of the apartment / Condominium
Area / 145m²
Outline / living room, dining room, kitchen, study room, master bedroom, master bathroom, master dressing room, bathroom
Main building material / oak, titanium-coated steel, polished porcelain tile, leather pattern melamine plates, wallpaper, tea mirror

02
01　03

01 _ 厨房与餐厅　02 _ 客厅　03 _ 餐厅与厨房
01 _ kitchen and dining room　02 _ living room　03 _ dining room and kitchen

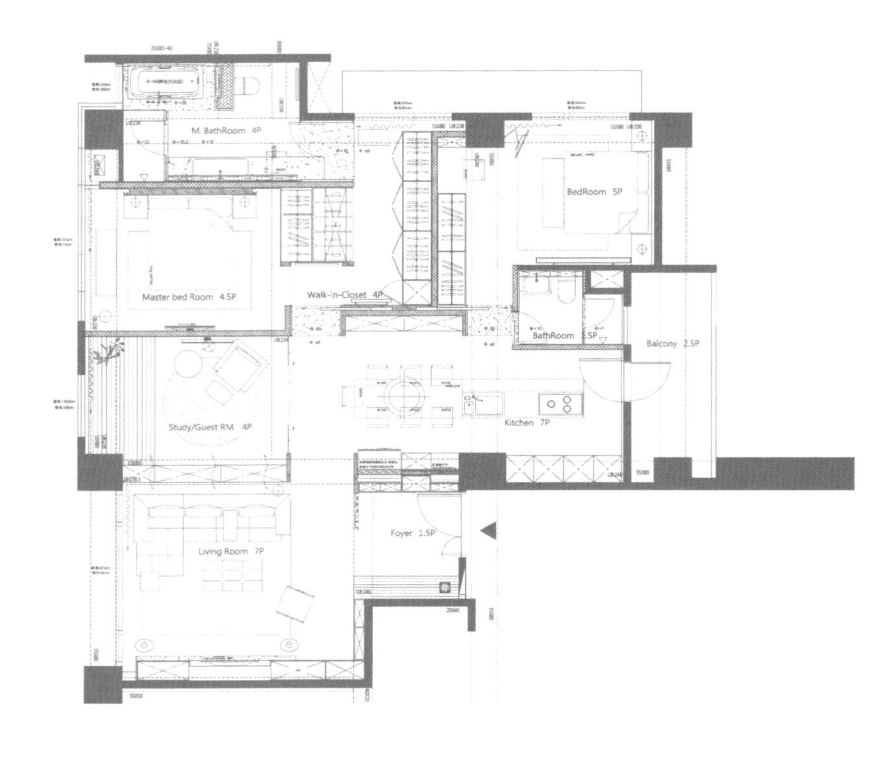

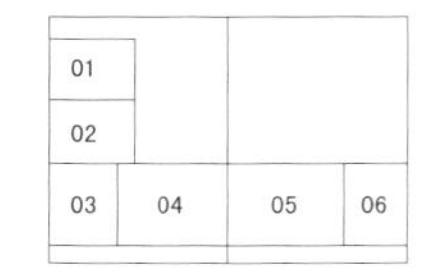

01 _ 主卧浴室　02 _ 主卧浴室　03 _ 主卧浴室入口
04 _ 主卧室　05 _ 书房　06 _ 玄关
01 _ master bathroom　02 _ master bathroom
03 _ master bathroom entrance　04 _ master bedroom
05 _ study room　06 _ entrance

Designer Data

赵玺
现　任／
大院室内装修有限公司主持人
学　历／
国立艺专雕塑组
经　历／
赖昌寿建筑师事务所
胡硕峰设计工作室
大匀设计（台北/上海）

Hsi Chao
Current Position
Supervisor of Da-Yuan Interior Design Ltd.
Education
Bachelor of Sculpture, National Taiwan University of Arts
Experiences
Lai-Chang-Shou Architectural Design Firm
Shry-Fong Hu Studio
W-design studio Taipei/Shanghai

柯元杰
现　任／
大院室内装修有限公司主持人
学　历／
实践大学空间设计系
东海大学建筑研究所
经　历／
佾所建筑研究室
奥代设计有限公司
胡硕峰建筑工作室
2007～2010 实践大学建筑系兼任讲师

Yuen-Chieh Ke
Current Position
Supervisor of Da-Yuan Interior Design Ltd.
Education
Bachelor of Architecture, Shih Chien University
Master of Architecture, Tunghai University
Experiences
Yi-Suo Architectural Studio
Ao-Dai Design Ltd.
Shyr-Fong Hu Studio
2007-2010 Adjunct Instructor of Department of Architecture, Shih Chien University

给生活一个简单纯净的空间

Provide your life a simple and pure environment

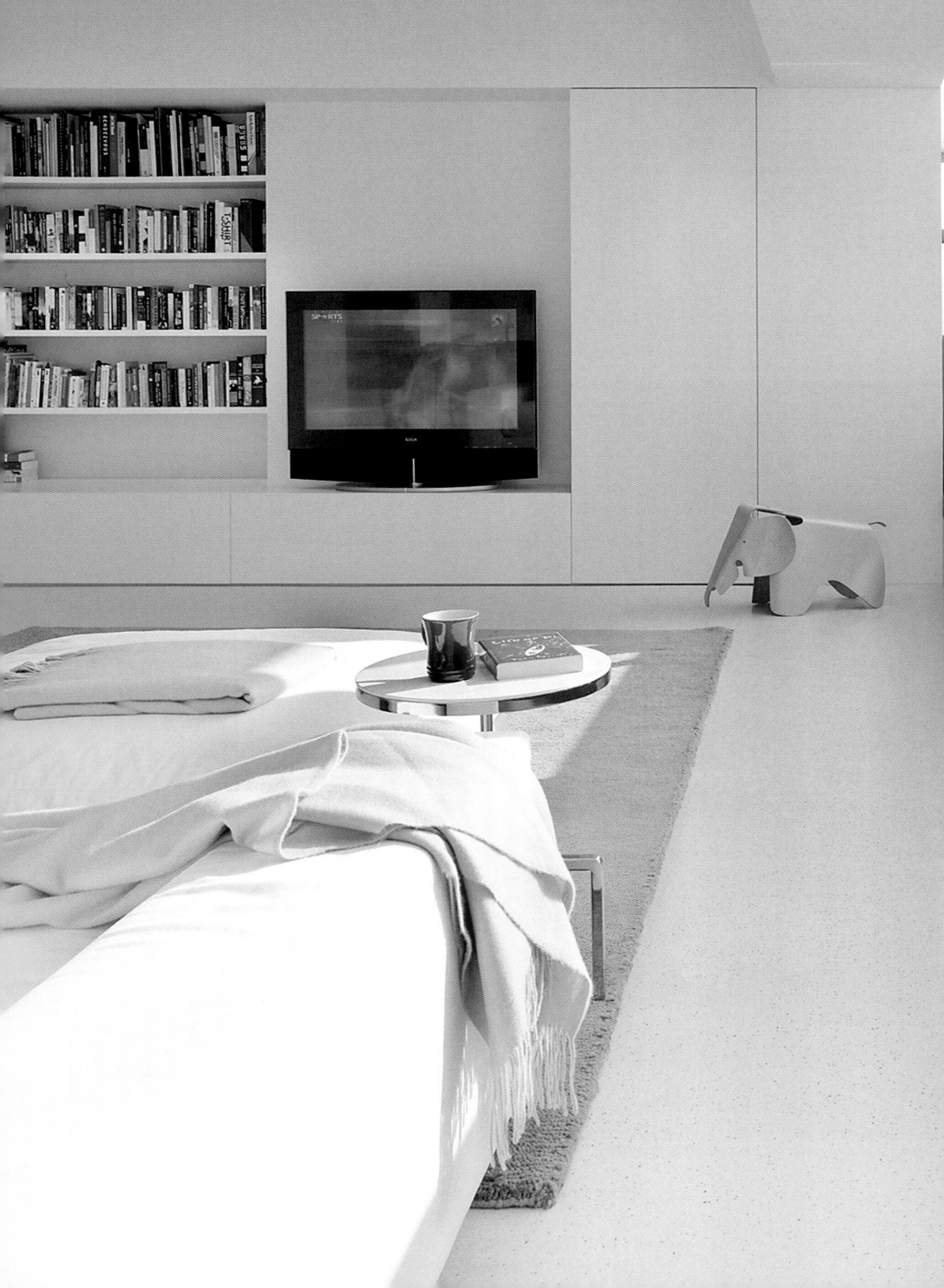

living room 客厅

living room 客厅

一个单身女子需要什么样的居住空间？通过与屋主的深入沟通，以及对都会生活的体验，设计师将盒子的意象融入设计，通过“盒中有盒”的概念，让私人空间被三面的公共空间所环绕，不复杂的层次与整体的格调一致，在清爽中见到一种纯真。

由于空间的格局方正，且三面窗户皆有景观，因此通过通透宽敞的设计让自然光线得以进入室内，搭配清爽的色调与利落的线条，创造出极具现代感的表情。将卧房包覆的手法，除了打造出方正的私人空间，隔断墙也正好界定出玄关。公共空间以L形呈现，受光线引领的视线更能感受到空间的大尺度。

在本案中，隐藏式设计将功能收整于立面，纯净的质感油然而生。客厅仅摆放一套简单的沙发，另在靠窗处配置座椅，为屋主提供另一个阅读与休憩的空间。厨房与餐厅采用开放式设计，通过中岛与餐桌的串联，消除建筑结构的阻碍感，造型组合柜体将功能性囊括其中，并赋予空间美感。

What kind of living space does a single woman want? By carefully analyzing the background of the female client as well as following the designer' s own cosmopolitan instinct, the concept of "box" is incorporated into the design. The idea in this design is that "there is a box within a box" . The private space is surrounded by three sides of the public zones. The use of uncomplicated layering is in accordance of the nature of this design: being pure in a refreshing environment.

In terms of the spatial layout, due to the fact that the apartment is rectangular in shape and all of its three windows provide a clear view to the scenery. Therefore, by the application of spacious design, a sufficient amount of light is allowed into the room. This goes well with light-colored paints and clean contour lines, which in turn creates a very modern look. Furthermore, not only a private space in a square shape is created by shielding the bedroom, its walls also helps to define an entrance. The public space is taken in a form of an "L" shape, which looks larger than it really is due to the abundance of lighting.

In this case, the concealed design allows functional features to be hidden under its clean outlook. A sense of purity can be suddenly felt. Only a sofa in minimalistic style and a seat next to the window are displayed at the living room, providing an alternative kind of resting and reading experience. Obstruction in the building structure is resolved through the use of "mid-island" that joins with the table, as well as the open design of the kitchen and the dining room. In any case, functionality is hidden within the stylistic design, providing the space a sense of visual beauty.

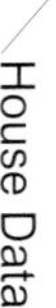

空间案名　敦南王宅
设 计 者　何侯设计
参与设计　何以立、侯贞夙、林于歆
空间地点　台北市大安区
空间性质　电梯大楼
空间面积　137 ㎡
空间格局　玄关、客厅、厨房、餐厅、客浴、主卧室、主卧更衣室、主卧浴室
主要建材　磨石地板、合成平光树酯地板、橡木集成材厨房台面、亚麻仁油环保板

Case Name / Dyna wangzhai
Designers / Yi-Li He, Chen-Su Hou, Yu-Hsin Lin
Company / Ho+Hou Studio Architects
Location / Da-an district, Taipei City
Nature of the apartment / Condominium
Area / 137m^2
Outline / entrance, living room, kitchen, dining room, guest's bathroom, master bedroom, master dressing room, master bathroom
Main building material / Pandomo terrazzo floors, plain synthetic resin floors, oak laminate kitchen countertops, linseed oil environmentally friendly board

01　03
02　04

01 _ 主卧室　02 _ 餐厅与厨房
03 _ 客浴与厨房　04 _ 主卧浴室
01 _ master bedroom　02 _ dining room and kitchen
03 _ guest's bathroom and kitchen　04 _ master bathroom

Designer Data

侯贞夙
现　任／
何侯设计 负责人
学　历／
1982～1985 台湾大学地理学士
1985～1988 哈佛大学景观建筑硕士
1989～1991 哥伦比亚大学建筑硕士
经　历／
1988～1989 Child Associates, 波士顿, 设计师。1991～1992 Rafael Vinoly Architects PC, 纽约设计师。1992～1993 Perkins & Will, 纽约设计师。1994～1997 Kohn Peterson Fox Associates PC,纽约设计师。1996 美国纽约州注册建筑师。1997～ Ho + Hou Studio Architects, 何侯设计 负责人。1997 淡江大学建筑系兼职讲师。

Chen–Su Hou
Current Position
Person in charge, Ho+Hou Studio Architects
Education
1982–1985 Bachelor of Geography, National Taiwan University
1985–1988 Master of Landscape Architecture, Harvard University
●1989–1991 Master of Architecture, Columbia University
Experiences
●1988–1989 Designer, Child Associates, Boston,
●1991–1992 Designer , Rafael Vinoly Architects PC, New York
●1992–1993 Designer, Perkins & Will, New York,
●1994–1997 Designer, Kohn Peterson Fox Associates PC,New York
●1996 Registered Architect in New York
●1997– Person in charge, Ho + Hou Studio Architects,
●1997 Adjunct Instructor of Department of Architecture, Tamkang University

Designer Data

何以立
现　任／
何侯设计 负责人
学　历／
1984 美国休斯敦大学建筑学士
1986 美国哈佛大学建筑系硕士
经　历／
1986 Raphael Moneo Architects, 波士顿，设计师。1987～1989 Schwartz/Silver Associate Architects Inc., 波士顿设计师。1991 美国马塞诸塞州注册建筑师。1989～1996 Pasanella + Klein Stolzman + &Berg 纽约, Associate Partner，协同负责人。1997～ Ho + Hou Studio Architects，何侯设计，负责人。1996～1998 实践大学空间设计系兼职讲师。1997~2000 东海大学建筑系兼职讲师。2001～2002 台湾交通大学建筑学院兼职讲师。2003～ 敬典文教基金会执行长。2009 台北科技大学兼职讲师。

Yi–Li He
Current Position
Person in charge, Ho+Hou Studio Architects
Education
1984 Bachelor of Architecture, University of Houston
1986 Master of Architecture, Landscape Architecture, and Urban Planning, Harvard University
Experiences
●1986 Designer, Raphael Moneo Architects, Boston
●1987～2000 Designer, Schwartz/Silver Associate Architects Inc., Boston
●1991 Registered Architect in Massachusetts
●1989–1996 Pasanella + Klein Stolzman + &Berg
New York , Associate Partner
●1997– Person in charge, Ho + Hou Studio Architects
●1996–2009 Adjunct Instructor of Department of Architecture, Shih Chen University
●1997–2000 Adjunct Instructor of Department of Architecture, Tunghai University
●2001–2002 Adjunct Instructor of Graduate Institute of Architecture, Taiwan Chiao Tung University
●2003– Chief Executive Officer of HOSS Foundation
●2009 Adjunct Instructor of National Taipei University of Technology

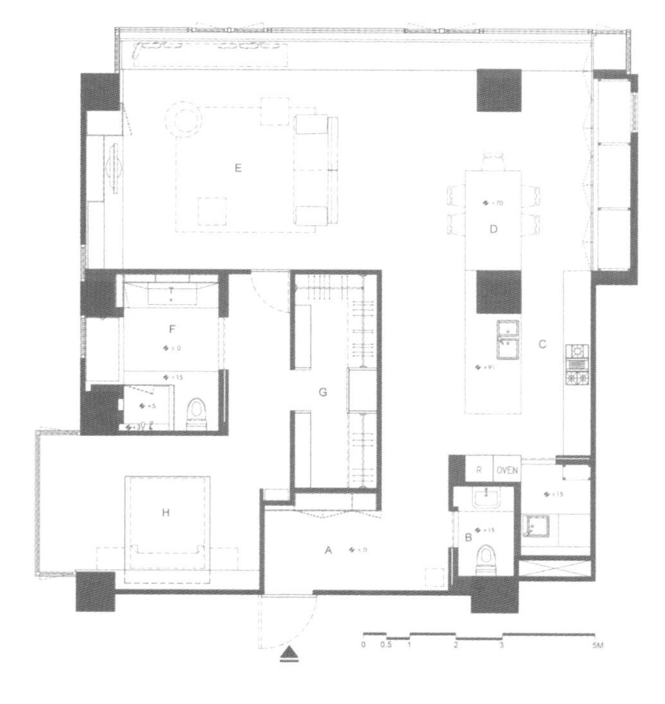

A - Entrance
B - Powder room
C - Kitchen
D - Working / Dining
E - Living room
F - Bathroom
G - Walk in Closet
H - Bedroom

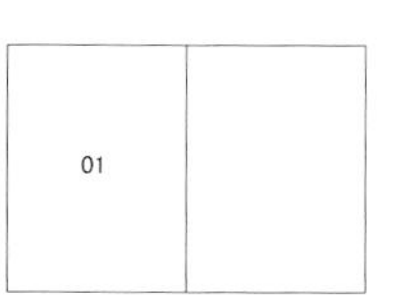

01＿主卧室　01＿master bedroom

舒缓安适让生活更美好

Comfort for a better life

Pictures_TWA Design
图片提供_传十设计

本案房屋四周充满绿色景物，为引入户外绿意，设计师不但将窗户改为落地窗，同时还运用"空间流动性"与"屋中盒"的概念，以几近全开放的手法，将全屋整合为一个大的开放空间。

借由无阻隔的动线设计，以及精巧的空间区隔，浴室、储藏室、衣橱等较需要封闭或遮挡的空间，不但能独立存在，而且仍整合为室内的一个"大盒"。因此，所有室内空间均可自由连通，畅行无阻。同时，窗外的绿意以带状方式围绕着室内，走到哪里皆可观赏到。

然而，在看似开放的空间中，仍保有了高度的使用自由度与弹性。活动拉门的设计，可以让空间变为封闭，让居住者保有一定程度的隐私。而在本案中，最特别的是主卧浴室，全然开放的设计，让居住者能有在山林中沐浴的感觉。

在风格设计上，整个空间以木质的质朴与温润，搭配深沉的灰黑色、灰蓝色，配合浅灰色的基底，并采取现代简洁融合人文的语汇，创造出一个休闲、舒适、沉静、与自然合一的美好居家。

This project is surrounded on all sides by a green landscape. In order to bring nature from the outside to the interior, the design makes use of picture windows that allow residents to gaze at the hillside in its entirety. Concepts of "spatial dynamism" and "box in a house" are adopted as well, using a near-total open style to merge the compartments of the entire house into a single, integrated and open space.

With the design of non-resistant flow dynamics and the clever use of space compartmentalization, spatial units such as the bathroom, storage room and closets that must be closed or covered can exist independently whilst remaining as an enclosure integral to the room. Hence, spaces within the room can be freely traversed without hindrance, and at the same time the greenery outside the windows can be brought about to surround the internal space, allowing residents a view of the extensive and rolling ecology surrounding their home anywhere within the house.

However, despite the seemingly complete openness, a high degree of functionality and flexibility is retained. Sets of sliding doors incorporated into the design can convert any open space into a closed compartment to provide the resident with a degree of privacy. The most novel design of this project is the open master bathroom with its completely opened design, allowing the resident the joys of a soothing bath within the wooded hills.

The design style utilizes the warm simplicity of wood with a deep tone of dark-gray; a sable gray-blue makes up the main theme and is complemented by a light gray base. A modern minimalist and human approach are taken to create a relaxing, comforting, becalming residence existing as one with the surrounding ecology.

House Data

空间案名　挹翠
设 计 者　李文心　许天贵
设计公司　传十室内设计有限公司
空间地点　台北市木栅
空间性质　公寓
空间面积　135 ㎡
空间格局　客厅、餐厅、厨房、书房、主卧室、客房、卫浴两间、阳台
主要建材　瓷砖、木材、玻璃、壁纸、窗帘、油漆

Case Name / Emerald Overflow
Designers / Wen-Hsin Li　Tien-Kui Hsu
Company / TWA Interior Design
Location / Taipei City, Mu-cha
Type of space / Residential Apartment
Area / 135 m^2
Layout / Living room, dining room, kitchen, master bedroom, guest room, two showering spaces, balcony
Main materials / Tiles, woodwork, glass, wallpaper, curtains, paint

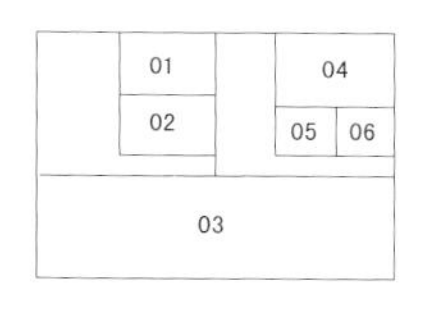

01 _ 客厅　02 _ 客厅与餐厅　03 _ 客厅与餐厅
04 _ 厨房、餐厅、客厅与书房　05 _ 餐厅与厨房
06 _ 书房
01 _ living room　02 _ living room and dining room
03 _ living room and dining room
04 _ kitchen, dining room, living room and study room
05 _ dining room and kitchen　06 _ study room

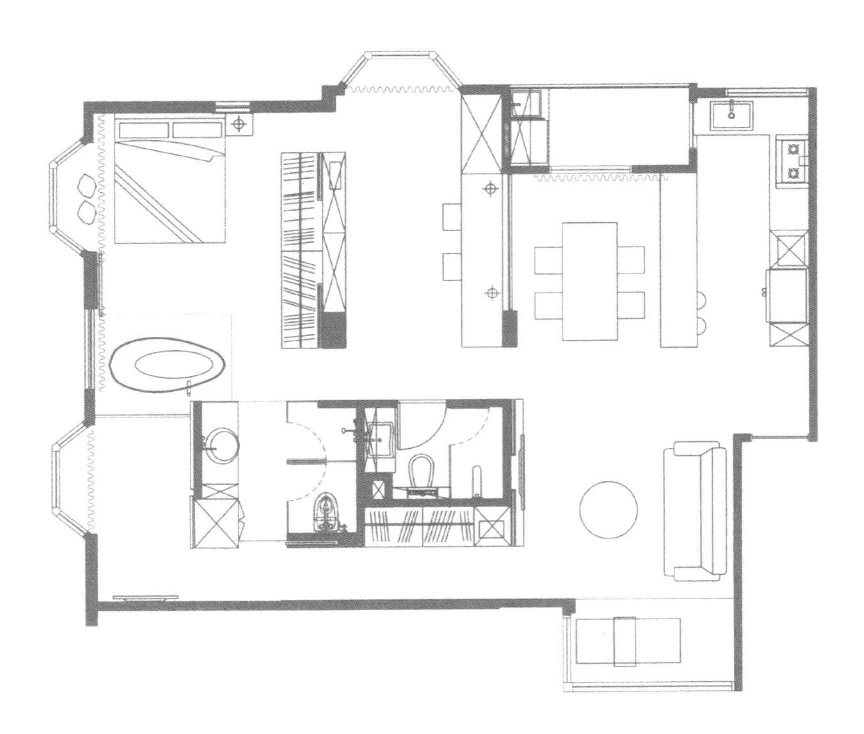

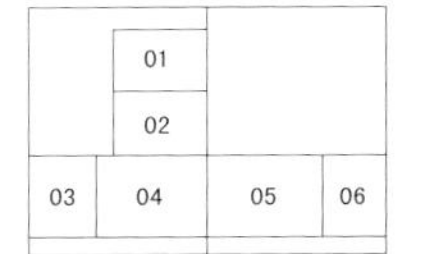

01 _ 主卧　02 _ 主卧　03 _ 主卧浴室
04 _ 主卧　05 _ 书房　06 _ 主卧浴室
01 _ master bedroom　02 _ master bedroom
03 _ master bathroom　04 _ master bedroom
05 _ study room　06 _ master bathroom

Designer Data

李文心
现　任／
传十设计设计总监
学　历／
台湾大学哲学系
美国新泽西州立大学社会学研究所
东海大学建筑硕士
经　历／
台湾室内设计协会理事
传十设计设计总监

Wen-Hsin Li
Education and Background
Department of Philosophy, National Taiwan University
Research Institute of the Department of Sociology, The State University of New Jersey
Masters, Department of Architecture, Tunghai University
Executive Member, Chinese Society of Interior Designers
Chief Designer of TWA Design

许天贵
现　任／
传十设计主持建筑师
学　历／
成功大学建筑研究所设计组硕士
经　历／
台北市建筑师公会会员
台湾室内设计协会会员
传十设计主持建筑师
许天贵建筑师事务所主持人

Tien-Kui Hsu
Education and Background
Masters, Research Institute of the Department of Architecture major of design, National Cheng Kung University
First place in the year 2000 National Architecture's Assessment, Architecture Category
Member, Taipei Architects Association
Member, Chinese Society of Interior Designers
Lead Architecture, TWA Design
Director, Tien-Kui Hsu Architecture and Design Firm

传十设计／
传十设计于2007年至2010年连续7次获得台湾室内设计大奖住宅类TID奖。成员皆具建筑专业背景。

TWA Design has consecutively won Taiwan Interior Design Award seven times from 2007-2010, with all the TID awards for the residential category. All our associates have a professional background in architecture.

living room客厅

环形动线串联自由空间

Circular moving line paints a free space

living room客厅

living room and dining room 客厅与餐厅

屋主从事金融业，在空间的设计上，设计师除了将铜币的意象融入于客厅之中外，空间也采用圆形的环绕动线。在本案中，书房设定为整个居家空间的中心，空间设计呈现出一种流畅而轻巧的韵律感。

设计师在整个空间的设计上，利用“环绕动线”的概念，从多功能书房、主卧及主浴室，通过动线的串联让空间关系更加紧密、顺畅。而在天花板的设计上，以“纸”的设计语汇铺展，从墙面近延伸至天花板的弧形转折处，营造出流畅而具雕刻感的趣味，呈现出立体的视觉效果。

在多功能的书房隔断上，使用了白色的圆形层叠的图案，象征着钱币堆积。另外，屋主希望所饲养的小狗也能在家中有属于它的活动范围，并可与私人区域区隔，因此设计师利用中空板和玻璃，结合了隐藏轨道，作为私人区域与公共区域的隔断，并可依照需求自由调整。整个空间采用动线与区域分隔的设计手法，加之大面积开窗，让室内更加明亮宽敞。

The owner has a finance background, so the designer embeds the idea of the coins in the living room, and creates a circular moving line for the space. The study is the center of the house. Therefore, smooth and ethereal rhythm is floating over the space.

The designer applies the concept of "circular moving line" to connect the multi-study, master bedroom, master bathroom in a smoother and more compact manner. For the design of the ceiling, the designer adopts the idea of paper: the wall extends to the ceiling and forms an arc turn, creating the smooth and sculpture-like delight and displaying three-dimensional visual effects.

Because the owner is in the finance industry, the designer uses white round totems piled together, to symbolize piles of coins. Additionally, as the owner would like to give his puppy a personal corner, the designer uses hallow board and glass plus hidden track to divide private and public areas, so that the owner can make adjustments when necessary. Overall, the designer's excellent moving line design along with space separation plus large windows make the whole space look more airy and shining.

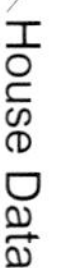

House Data

空间案名 植心园
设 计 者 陆希杰
参与设计 曹车政、张琼之
设计公司 陆希杰设计事业有限公司
空间地点 台北
空间性质 电梯大楼
空间面积 133 m²
空间格局 客厅、厨房、视听室
主要建材 瓷砖、烤漆铁板、印度黑花岗石、染黑橡木、清玻璃、阳光卷帘

Title / The National Garden
Designer / Shi-chieh Lu
Assistant designer / Che-zheng Cao, Qiong-zhi Zhang
Design company / CJ STUDIO
Location / Taipei City
Category / Elevator building
Area / 133 m²
Layout / Living room, kitchen, and audio video room
Building material / tiles, iron paint, Indian black granite, black oak, clear glass, sunshine roller shutters

01 02	04
03	05

01 _ 客厅与餐厅 02 _ 餐厅 03 _ 客厅与餐厅 04 _ 客厅、餐厅与厨房 05 _ 书房
01 _ living room and dining room 02 _ dining room 03 _ living room and dining room 04 _ living room and dining room and kitchen 05 _ study room

01
06
02 03 04 05 07

01 _ 主卧室　02 _ 书房　03 _ 主卧室与主卧浴室　04 _ 走道　05 _ 浴室　06 _ 主卧浴室　07 _ 主卧浴室
01 _ master bedroom　02 _ study room　03 _ master bedroom、master bathroom
04 _ aisle　05 _ bathroom　06 _ master bathroom　07 _ master bathroom

陆希杰
现　任/
CJ STUDIO 主持人
学　历/
东海大学建筑系毕业。英国AA建筑联盟硕士
简　介/
从事建筑及室内设计、家具设计、产品设计以及相关领域的研发，现兼任交通大学建筑研究所助理教授，著有《锻造视差》。代表作品有："窦腾璜及张李玉菁wum 概念店"；"窦腾璜及张李玉菁台中Tiger city概念店"（2005年获得日本JCD设计大奖）；"Aesop微风店"（2006年获得日本JCD设计大奖、国际室内设计联盟IFI 2007设计金奖）；"敦南国美蔡宅作品"（获得台湾室内设计大奖TID Award 2007金奖）。入选美国室内设计杂志《INTERIOR DESIGN七十五周年特集》中五位具潜力设计师之一。2008年，参展"第十一届威尼斯建筑双年展"和台北当代艺术馆"第六届城市行动艺术节"。

Shi-chieh Lu
Professional
CJ STUDIO, design director
Education
Master of Architectural Association School of Architecture, UK
Department of Architecture, Tunghai University
Brief Bio
Shichieh Lu is an adjunct assistant professor in Graduate Institute of Architecture, NTCU, specializing in interior design, furniture design and product design. He authors
Forging parallax, published in 2003. Lu's masterpieces include
•Stephanie Dou & Changlee Yugin's concept store, WUM
•Stephan Dou & Changlee Yugin's concept store in Taichung, Tiger City (won 2005 JCD Design Award)
•Aesop in Breeze Store (won 2006 JCD design award and IFI 2007 golden prize)
•Cai's house in Dun-nan Guo-mei (won TID Award 2007 golden prize) .
Lu, listed as one of the five promising architects in Interior Design Magazine's special issue for its 75 anniversary, participated in the 2008 Biennale Architecture 11th International and 2008 MOCA TAIPEI-dark city

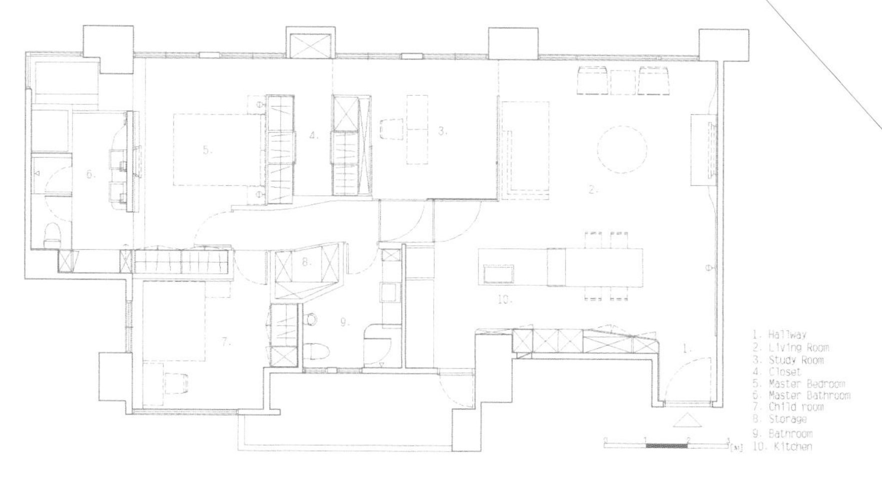

living room 客厅

在家也能感受巴黎的时尚与优雅

Let your home to be as modern and elegant as Paris

living room 客厅

优雅可以有多种方法诠释，在本案中，每一个角落都能自成一道优美的风景。通过设计师细腻的营造与安排，回到家，竟也能转换时空般地跳跃到巴黎，感受到属于巴黎都会的细致与高雅，还有融合异国文化的多元情调。

玄关，通过精雕细琢的屏风构成。屏风区隔了空间，也让空间通透，同时更成为空间中视觉造型的一部分，独特的姿态就先以其细腻的表现展现出层次感。客厅的设计结合大面积的落地窗，构成一幅美丽的风景。中式家具的使用，将法式的优雅和东方的人文情怀充分融合，让东西方美学通过空间设计在此交锋，展现出主人独特的品位。

空间的串联也呈现多元化，餐厅为整个居家空间的重心，公共空间与私人空间正好以此为中心成辐射分布。私人空间通过同样的法式设计语汇展现出优雅的气质，营造休憩空间的私密和甜美。色调的使用也巧妙地与壁纸相呼应，让空间不但具有多层次的美感，也展现出均衡的一致性。

There are many ways you can interpret elegance, but in this case, every corner of the space are self-contained with beautiful scenery. Through the careful planning by the designer, an illusion of being at Paris is created when one walks into this property. At this place, you can feel the gracefulness and sophistication as belonging to the urbane Paris, as well as immersing yourself into the atmosphere of being surrounded by this charming foreign culture.

The appearance of the entrance is shaped by an exquisitely carved screen. The screen not only acts as a divider but also allows the penetration of the space, and at the same time, represents an important part of the visual design. Its uniqueness can be seen from the intricately exhibited layering effect. A large window is positioned at the living room, which provides a view to the beautiful scenery. The use of Chinese-styled furniture allows the fusion of French elegance and Oriental sentimentality. Anyhow, Eastern and Western aesthetics are met here, fully exhibiting the unique taste of the house owner.

In terms of the spatial connectivity, a multi-dimensional presentation is also used. The dining room acts as the central point of the entire home environment. Both the public and the private space radiate from this point. Again, the private space is designed in the French style, which is both elegant and comfortable. Furthermore, the wallpapers also use similar colors/tones, which not only provide a sense of multi-layered beauty, but also display a symmetrical splendor.

Pictures_Simple Interior Design
图片提供_尚展设计

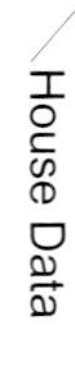

空间案名	刘宅
设 计 者	吴启民
设计公司	尚展设计
空间地点	台北市
空间性质	电梯大楼
空间面积	132m²
空间格局	玄关、客厅、厨房、餐厅、书房、主卧室、主卧更衣室、主卧浴室、小孩房、客浴
主要建材	进口超耐磨地板、进口壁纸、铁件、激光雕花

Case Name / Liu's apartment
Designer / Chi-Min Wu
Company / Simple Interior Design
Location / Taipei city
Nature of the apartment / Condominium
Area / $132m^2$
Outline / entrance, living room, kitchen, dining room, study room, master bedroom, master dressing room, master bathroom, kid's room, guest's bathroom
Main building material / imported hard-wearing floor, imported wallpaper, metal components, laser-cut decoration

01 02	04
03	05

01＿玄关 02＿餐厅 03＿餐厅 04＿餐厅 05＿餐厅
01＿ entrance 02＿ dining room 03＿ dining room
04＿ dining room 05＿ dining room

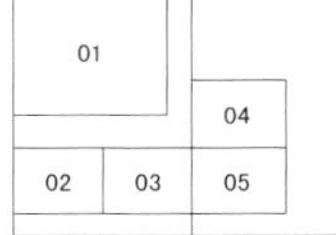

01 _ 小孩房　02 _ 主卧室　03 _ 主卧室　04 _ 主卧浴室
01 _ kid's room　02 _ master bedroom
03 _ master bedroom　04 _ master bathroom

Designer Data

吴启民
现　任/
尚展设计 设计总监
学　历/
加拿大英属哥伦比亚大学（UBC）建筑与都市计划学系毕业
经　历/
汇侨设计、展亦设计 设计总监
2008年荣获AID Award 台湾室内设计大奖
2009年AID Award 台湾室内设计大奖入围
简　介/
始终坚持要利用房子本身的特性规划，顺着房子的优点来设计。有形的功能，回归自然变得纯粹。无形的价值，就不会被时间的潮流所淹没。奢华最极致的表现是融合自然与科技。着墨于每个微小的细节，让居住者的身心都能获得无比的满足与自在，这正是当前品味之士所追求的。

Chi-Min Wu
Current Position
Chief Design Officer of Simple Interior Design
Education
Bachelor of Architecture and Landscape Architecture, The University of British Columbia
Experiences
Chief Design Officer of Rich Honour Design Group and United Design Associate,
2008 AID Award
2009 Nominated AID Award
Brief
When designing for an apartment, I always strive to take the advantage of the original characteristic of the house. I insist to keep the good points intact. Functionality with forms allows the nature to become purer, so that values without forms would not disappear amongst the tide of time. The ultimate luxury is to combine technology and nature. I mind every single detail, so that both the mind and body of my clients can be satisfied and contended, the two values are what people with good taste would pursue.

反转，勾勒大气的生活场域

Reverse it! Have a brand new lifestyle

panoramic view of the living room and second floor lounge entrance 客厅挑高与二楼起居室全貌

living room 客厅

在规则中舍弃通常的手法，以反向的概念进行设计，是本案的独特之处。在此跃层空间中，设计师一反将一楼设定为公共区域的做法，将卧房配置于此处，厨房与餐厅则移至二楼。通过空间的迁移，居住者既有的生活习惯被改变了，为生活创造出更多的可能。

打破框架的设计手法，让人一进门即有深刻感受。楼梯的斜线正好成为玄关与客厅的隔断，也区分了客厅与两个主卧空间。客厅区经由阶梯的导引而进入，大量的光线与突然放大的水平、垂直向度，带来无比宽敞的感受。地面的层次变化自然地界分空间，也让空间尺度得到最大的释放。而将餐厅与厨房设定于二楼的配置，也打破了以往的生活秩序，替居住者创造另一种新的生活形态。

本案另一个精彩的空间在主浴室。烤漆面板、木皮与瓷砖，同色系但不同材质的装饰元素富有层次又和谐，仿佛在共同演奏一首精彩的交响曲。通过对空间的重新思考，设计师让大宅有了全新的演绎方式，勾勒出不一样的生活面貌。

Standard rules of practice are abandoned in the designing of this apartment, and are replaced with a new concept of reversal, the thinking is what makes this property unique. In the designing of this duplex apartment, the designer ditches the commonly held wisdom of assigning the first floor as the public space. Instead, bed room is assigned here. As for the kitchen and dining room, they are moved to the second floor. Through this spatial transformation, the existing habits of the house owner are also changed, creating more possibilities in life.

The practice of breaking the normal structural frames can be felt as soon as one walks into the apartment. The diagonal line of the staircase also acts as a divider between the entrance and the living room, in the same time partitioned the two master bedrooms from the living room. As guided through by the ladder into the living room area, an ample amount of light can be suddenly felt. This visually extends the horizontal and vertical dimensions, creating an illusion of an enlarged space. Furthermore, the variations in the floor height also create natural boundaries for the living space. This clever designing technique also brings a feeling of a larger space. As for allocating the dining room and kitchen to the second floor, this also brings about a reversal to the standard order of life for the house owner as well as creating a new lifestyle.

Another interesting aspect of this design lies in the main bathroom; by using different materials with the same color and tonality, whether they are enamel paint, woodwork and tiles, all brings about an appearance that is harmonious yet rich in layering in the same time, just like a well-conducted symphony. Through the use of unconventional spatial design, the designer brings about a brand new face to this apartment.

House Data

设 计 者　何侯设计
参与设计　何以立、侯贞夙、钟静嘉、杨雅婷、连国伸、周正峰、刘懿萱
空间地点　台北市大直区
空间性质　跃层
空间面积　120 ㎡
空间格局　玄关、客厅、餐厅、厨房、起居室、主卧室×2、主卧更衣室、主卧浴室×2、客浴、客房
主要建材　集成石材、木皮、背漆镜子、环保板、水泥地板

Company / Ho+Hou Studio Architects
Designers / Yi-Li He, Chen-Su Hou, Ching-Chia Chung, Ya-Ting Yang, Kuo-Shen Lien, Cheng-Feng Chou, Yi-Hsuan Liu
Location / Xinyi District, Taipei City
Nature of the apartment/ duplex apartment
Area / 120m^2
Outline / entrance, living room, dining room, kitchen, lounge, master bedroom x2, master dressing room, master bathroom x2, guest's bathroom, guest's room
Main building material / integrated stone works, wood, back-painted mirror, Farbo environmentally friendly board, cement floor

01 _ 客厅　02 _ 主卧浴室　03 _ 玄关
01 _ living room　02 _ master bathroom
03 _ entrance

second floor lounge 二楼起居室

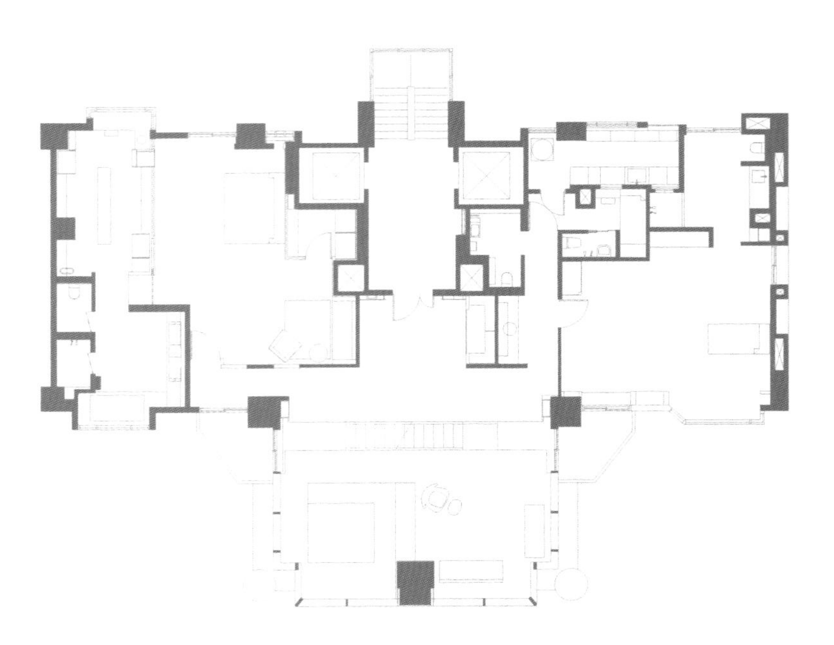

侯贞夙
现　任/
何侯设计 负责人
学　历/
1982~1985 台湾大学地理学士
1985~1988 哈佛大学景观建筑硕士
1989~1991 哥伦比亚大学建筑硕士
经　历/
1988~1989 Child Associates，波士顿，设计师。1991~1992 Rafael Vinoly Architects PC，纽约，设计师。1992~1993 Perkins & Will，纽约，设计师。1994~1997 Kohn Peterson Fox Associates PC,纽约，设计师。1996 美国纽约州注册建筑师。1997~ Ho + Hou Studio Architects，何侯设计负责人。1997淡江大学建筑系兼职讲师。

Chen-Su Hou
Current Position
Person in charge, Ho+Hou Studio Architects Education
1982-1985 Bachelor of Geography, National Taiwan University
1985-1988 Master of Landscape Architecture, Harvard University
•1989-1991 Master of Architecture, Columbia University
Experiences
•1988-1989 Designer, Child Associates, Boston,
•1991-1992 Designer , Rafael Vinoly Architects PC, New York
•1992-1993 Designer, Perkins & Will, New York,
•1994-1997 Designer, Kohn Peterson Fox Associates PC,New York
•1996 Registered Architect in New York
•1997- Person in charge, Ho + Hou Studio Architects,
•1997 Adjunct Instructor of Department of Architecture, Tamkang University

Designer Data

何以立
现　任/
何侯设计 负责人
学　历/
1984 美国休斯敦大学建筑学士
1986 美国哈佛大学建筑系硕士
经　历/
1986 Raphael Moneo Architects，波士顿，设计师。1987~1989 Schwartz/Silver Associate Architects Inc.,波士顿，设计师。1991 美国马塞诸塞州注册建筑师。1989~1996 Pasanella + Klein Stolzman + &Berg 纽约，Associate Partner，协同负责人。1997~ Ho + Hou Studio Architects，何侯设计，负责人。1996~1998 实践大学空间设计系兼职讲师。1997,2000 东海大学建筑系兼职讲师。2001~2002 台湾交通大学建筑学院兼职讲师。2003~ 敬典文教基金会执行长。 2009 台北科技大学兼职讲师。

Yi-Li He
Current Position
Person in charge, Ho+Hou Studio Architects
Education
1984 Bachelor of Architecture, University of Houston
1986 Master of Architecture, Landscape Archi-tecture, and Urban Planning, Harvard University
Experiences
•1986 Designer, Raphael Moneo Architects, Boston
•1987-2000 Designer, Schwartz/Silver Associate Architects Inc., Boston
•1991 Registered Architect in Massachusetts
•1989-1996 Pasanella + Klein Stolzman + &Berg
New York , Associate Partner
•1997- Person in charge, Ho + Hou Studio Architects
•1996-2009 Adjunct Instructor of Department of Architecture, Shih Chen University
•1997,2000 Adjunct Instructor of Department of Architecture, Tunghai University
•2001-2002 Adjunct Instructor of Graduate Institute of Architecture, Taiwan Chiao Tung University
•2003- Chief Executive Officer of HOSS Foundation
•2009 Adjunct Instructor of National Taipei University of Technology

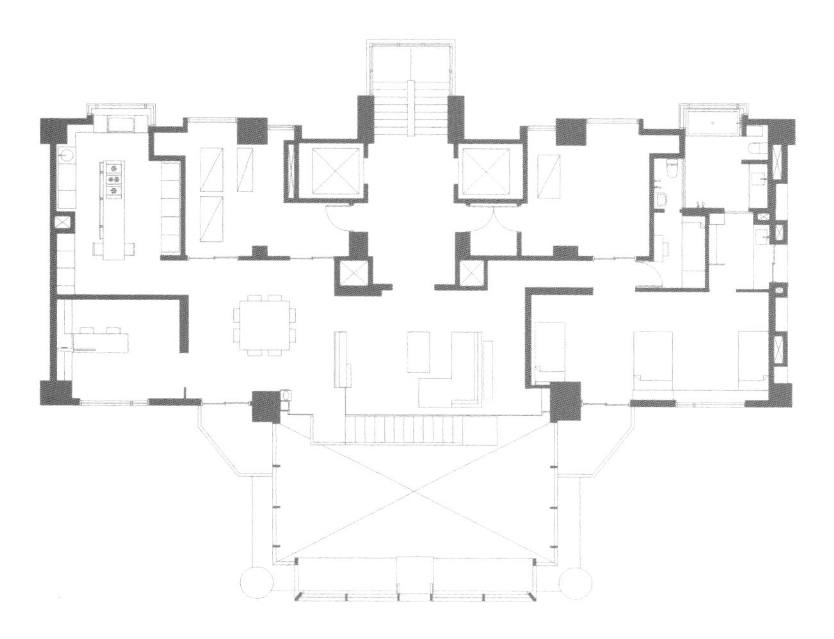

对生活空间做一个美好的实验

Experiment with your life

dining room 餐厅

Pictures_ Materiality Design Photographer_ Kyle Yu
图片提供_石坊空间设计研究事务所 摄影_Kyle Yu

屋主为年轻艺术家，专于音乐与绘画。依循居住者的背景与生活习惯，在本案设计中融合了艺术气息、都会的休闲风格以及展演空间的概念。通过活动的墙面设计，将展演空间加以延伸，也将居家打造成具有丰富意涵的场所。

空间的设计讲求自由无阻隔，因此多余的墙面均被拆除，空间通过地面切割区分，客厅、琴房以及主卧室地面铺以木地板，而餐厅、厨房、客浴以及次卧房地面则使用水泥粉光，表现出粗犷与年轻的气息。除此之外，琴房与客厅间利用电视墙区隔，凸显出琴房的演奏厅意象，活动拉门则可让两个空间既连接又各自独立。

画廊的概念亦融入空间，设计师在玄关入口处的长廊结合展示空间的概念，摆放屋主的画作。中性的灰色水泥正好衬托出作品的个性，也让空间感更为稳定。次卧房采用半开放式手法设计，拉门可灵活移动，展现出无阻隔的空间格调。

本案空间具有极高的变化性，可依居住者的使用习惯而时时展现出不同的表情，让居家个性因而更为鲜活。

The background of the house-owner is a young artist, who studied music and painting. In accordance to the background and lifestyle of the occupant, this design incorporates artistic elements, Urban Loft style as well as the concept of space extension. The use of movable wall design gives an illusion of an extended space, as well as making the apartment a stylish place to live in.

Since the design concept strives for a smooth and refined effect, excessive walls are removed. The partitioning of space is achieved by using different flooring materials; wood flooring is used on the living room, the piano room as well as on the master bedroom, whereas polished concrete flooring, which gives off the vibe of ruggedness and youth, is used on the dining room, kitchen, guest's shower room, and bedroom. In addition, television wall is used as a divider to separate the piano room from the living room. This design also highlights the auditorium-like image of the piano room. The use of sliding doors also allowing the two rooms to be independent from each other.

Furthermore, the concept of having an art gallery is also incorporated into the design. Along the long corridor facing the entrance, paintings of the house- owner are displayed. The use of grey cement (a neutral color) on the walls brings out colors and personality to the paintings and, at the same time, its calm and subdued coloration also give the place a matured appearance.

Sliding doors are used on the second bedroom, rendering the room semi-opened, thus emphasizing the unobstructed nature of the living space. In summary, the design of this property exhibits great versatility: the apartment constantly changes its look depending on the way the inhabitant uses it; this in turn adds a lot of personality to the home.

House Data

空间案名　怀想美好
设 计 者　郭宗翰
参与设计　张丰祥
设计公司　石坊空间设计研究
空间地点　台北市
空间性质　电梯大楼
空间面积　118 m²
空间格局　客厅、餐厅、厨房、琴房、主卧室、次卧室、主浴室、客浴室
主要建材　石材、玻璃、水泥、染色木皮、黑铁件、实木、特殊涂料

Case Name / Remember the good times
Designer / Tsung-Han Kuo
Co-designer / Feng-Hsiang Chang
Company / Materiality Design
Location / Taipei City
Nature of the apartment / Condominium
Area / 118m²
Outline / living room, dining room, kitchen, piano room, master bedroom, second bedroom, master bathroom, guest's bathroom
Main building material / stone, glass, cement, stained wood, black iron, hardwood, special paint

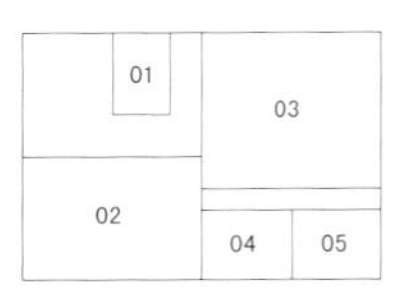

01 _ 玄关　02 _ 餐厅与客厅　03 _ 餐厅与客厅　04 _ 客厅　05 _ 琴房
01 _ entrance　02 _ dining room and living room
03 _ dining room and living room　04 _ living room　05 _ piano room

Designer Data

郭宗翰

现　任/

石坊空间设计研究 设计总监

学　历/

1997~1999　英国伦敦艺术大学空间设计系学士

1999~2000　英国北伦敦大学建筑设计系硕士

经　历/

2000~2002　香港商穆氏设计　设计师

2002　石坊空间设计研究　设计总监

2005　实践大学设计学院兼职讲师

Tsung-Han Kuo

Current Position

Director, Materiality Design

Education

1997-1999　University of the Arts London, BA in Spatial Design

1999-2000　University of North London, UK , MA in architectural design

Experiences

2000-2002　Designer , Moody's Commercial Design, Hong Kong

2002　Director, Materiality Design

2005　Lecturer at College of Design, Shih Chien University

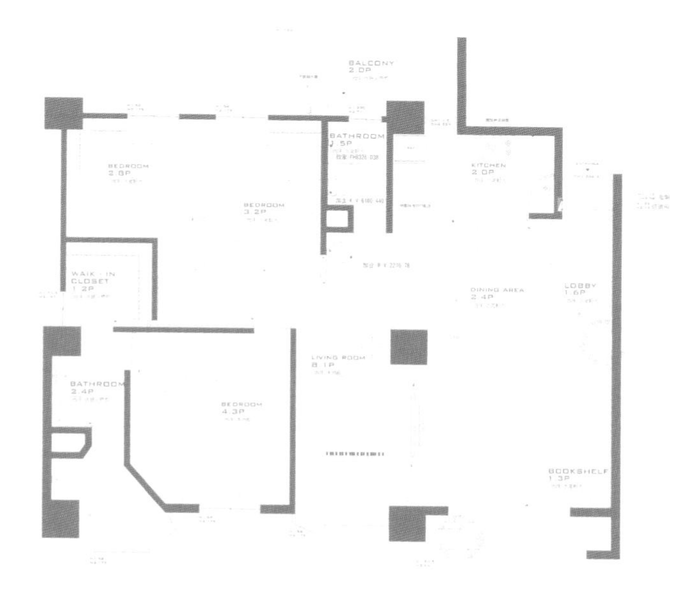

01	
02	03 / 04 / 05

01＿次卧室　02＿次卧室　03＿主卧室　04＿次卧室　05＿走道

01＿second bedroom　02＿second bedroom　03＿master bedroom

04＿second bedroom　05＿aisle

图书在版编目（CIP）数据

极致大宅/台湾麦浩斯公司编.—福州：福建科学技术出版社，2012.5
ISBN 978-7-5335-4000-5

Ⅰ.①极… Ⅱ.①台… Ⅲ.①住宅-建筑设计-作品集-台湾省-现代 Ⅳ.①TU241

中国版本图书馆CIP数据核字（2011）第272304号

书　　名　极致大宅
编　　者　台湾麦浩斯公司
出版发行　海峡出版发行集团
　　　　　福建科学技术出版社
社　　址　福州市东水路76号（邮编350001）
网　　址　www.fjstp.com
经　　销　福建新华发行（集团）有限责任公司
排　　版　福建科学技术出版社排版室
印　　刷　中华商务联合印刷（广东）有限公司
开　　本　635毫米×965毫米　1/8
印　　张　38
图　　文　304码
版　　次　2012年5月第1版
印　　次　2012年5月第1次印刷
书　　号　ISBN 978-7-5335-4000-5
定　　价　338.00元